Economic Parables & Policies

An Introduction to Economics

Third Edition

Laurence S. Seidman

M.E.Sharpe
Armonk, New York
London, England

Library of Congress Cataloging-in-Publication Data

Seidman, Laurence S.
    Economic parables & policies : an introduction to economics / Laurence S. Seidman.—
3rd ed.
        p. cm.
    Includes index.
    ISBN 0-7656-1108-2 (hardcover)
        1. Economics. 2. United States—Economic policy. I. Title: Economic parables and
polities. II. Title.

HB171.5.S454 2004
330–dc22

                                                                                    2003056716

Printed in the United States of America

The paper used in this publication meets the minimum requirements of
American National Standard for Information Sciences
Permanence of Paper for Printed Library Materials,
ANSI Z 39.48-1984.

BM (c)   10   9   8   7   6   5   4   3   2   1

For my parents, Eleanor and Irving Seidman

Contents

Tables and Figures

Preface

The purpose of this book is to provide a lively and enjoyable one-semester introduction to economics. Can good economics be entertaining—at least some of the time? Is it possible to learn economics while smiling—at least occasionally? Can economic policies be made interesting—at least most of the time? I believe the answer to these questions is yes. That's why I've written this introduction to economics. For many years, I've been using the material in this book to teach a one-semester course at the University of Delaware.

What's different about this introduction to economics? After all, there are quite a few introductory economics books around. Several things. First, it uses parables and a lively cast of characters to make economics enjoyable and memorable: Adam and Eve (chapter 1), the passionate objector (chapter 3), Ricardo (chapter 4), Senator Myopia and Old Karl (chapter 7), emaciated S and the lazy heir (chapter 8), the Productives and the Tryers (chapter 9), the senators of the Aroman Food Crisis (chapter 11), and the extraterrestrial XT (chapter 12). Second, it focuses on economic policy issues: the virtues of free markets and the shortcomings of price controls, environmental pollution, the case for free trade, monetary and fiscal policies to combat recession and inflation, tax reform to promote economic growth, social security, health insurance, education, and poverty. Third, it teaches two and only two basic diagrams—the production possibilities curve and demand and supply—and uses them to teach basic microeconomics and basic macroeconomics, and to analyze policy issues. In contrast to quite a few introductory books, it doesn't use up a lot of time and energy mastering cost

curves, so it has enough time and energy left to treat important policy issues.

First and foremost, this book is written for the college student who will take only one semester of economics. In writing this book, and teaching this course, I have asked myself: What are the most interesting, important, and useful things I can teach about economics in a single semester? I tell my students that I intend, in one semester, to give them "economics to last a lifetime." But second, this book is for students who initially intend to take only one semester of economics, but soon realize that economics is fun, interesting, and important, and decide to go on and take at least a semester of microeconomics and a semester of macroeconomics. By making their introduction to economics as interesting as possible, I hope to persuade students to take more economics.

And now, without further delay, let's turn to our first parable, entitled "An Economist's Genesis."

Economic Parables & Policies

I

Introduction to Economics

1 | An Economist's Genesis

In the beginning, Adam and Eve had no tools. To compensate, God saw to it that the weather and soil never failed them. Initially, Adam and Eve devoted all their working time to growing food. With their bare hands, they plowed, planted, and harvested. Each year, they consumed all the food they produced, and each year, production and consumption remained the same. For all we know, Adam and Eve were happy.

Adam's Dream

But one night, Adam had a dream. With an imagination that leaped centuries, Adam dreamed of a tractor. In the dream, Adam saw their ability to plow, plant, and harvest multiply.

"If only we had a tractor." The thought haunted Adam and Eve for weeks as they continued to farm with their bare hands. Naturally, they prayed daily to God to give them a tractor. But to no avail.

One day, Adam and Eve were walking in the garden.

"So you want a tractor?" spoke a voice. And they knew the voice was God's.

"Yes," they replied, trembling. And God answered, "Do you expect a tractor to fall from the sky like manna from Heaven?"

And Adam and Eve whispered, "That's exactly what we were hoping."

Then God laughed, and in a kind voice said, "No, my children, it is time for you to eat from the tree of economic knowledge. I will not give you a tractor. You must make your own tractor, by the sweat of your brow."

"That's just what we were afraid of," said Adam and Eve. Then God burned detailed instructions—"How to Make a Tractor"—onto a tablet of stone that lay at the foot of the tree of economic knowledge. As they sat under the tree, eating its fruit and reading the instructions, Adam and Eve suddenly realized a fundamental truth of the human condition. They realized that, while making a tractor, they must devote less time to plowing, planting, and harvesting food. In the short run, they must reduce their consumption of food.

"We face a trade-off," said Eve suddenly. "We must sacrifice consumption in the present, while we build the tractor, in order to enjoy more consumption in the future."

"Isn't there any way around this?" asked Adam gloomily.

"I'm afraid not," said Eve, who may have been the second person, but was clearly the first economist. "No sacrifice, no rise in the standard of living. It's that simple, honey."

Opportunity Cost and the Production Possibilities Curve

"Could you show me this trade-off in a simple diagram?" asked Adam. "I'm better at pictures than words."

"Of course," said Eve. "Let me draw Figure 1.1 in the dirt with this stick. On the horizontal axis is the number of units of food, and on the vertical axis, the number of tractors. If we devote all our time to producing food this year, we can produce 100 units, so one option for us is point A: produce 100 units of food and no tractors. I estimate that we can make 1 tractor if we devote 40 percent of our time to building a tractor and only 60 percent of our time to producing food, so another option for us this year is point B: produce 60 units of food and build 1 tractor. I estimate that another option is point C: produce only 20 units of food and build 2 tractors."

"So what is the *cost* of building a tractor?" asked Adam.

"Why, it's 40 units of food. To build 1 tractor, we must forego the opportunity of producing 40 units of food, so that's the *opportunity cost* of 1 tractor."

Figure 1.1 **Production Possibilities Curve**

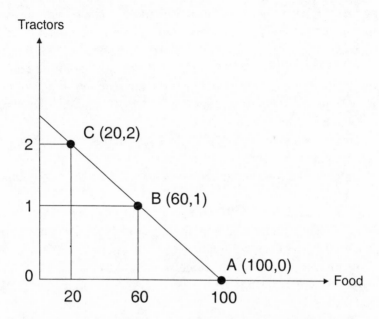

"Is there a name for the line in Figure 1.1?" asked Adam.

"I'll give it one," said Eve. "I'll call it the *production possibilities curve*, because it shows the possible combinations of food and tractors that we can produce this year."

"But why do you call it a *curve*?" asked Adam. "It looks like a line to me."

"It will be a straight line," replied Eve, "if making another tractor always requires giving up exactly 40 units of food. But maybe making the second tractor requires giving up more than 40 units, or less than 40 units; if so, then it would be a curve, not a straight line. To keep it simple, I'll assume a straight line; but to be ready for other outcomes, I'll call it a *production possibilities curve*."

Accounting: Output (GDP), Income, Consumption, Investment, and Saving

Having eaten from the tree of economic knowledge, Adam and Eve decided to do their accounting properly. Although they knew it was

unnecessary in their simple economy, they decided it would be fun to use money to keep track and, more important, to help their descendants understand the fundamentals. Conveniently, identical rectangular-shaped green leaves hung from the branches of the tree of economic knowledge. By fiat, Adam and Eve declared each leaf to be one dollar ($1). Here's how they did their accounting.

They began with the pre-tractor economy. Each year they produced a hundred units of food.

"*Output* is what we produce," said Eve, "and *inputs*—labor and land—are what we use to produce output."

"Inputs go into the production process," said Adam, "and output comes out."

"Exactly," replied Eve. "If we had a tractor, then we would have another input—capital—and with three inputs, labor, land, and capital, we could produce more output."

They arbitrarily declared the price per unit of food to be $1, so that total output (which they also called *gross domestic product*, or *GDP*) was $100.

To illustrate a basic truth of national income accounting, Adam and Eve engaged in a ritual. Playing the role of consumers, they paid $100 for the food, which they called *output*, placing the hundred leaves on a flat boulder beneath the tree of economic knowledge. Immediately, they circled to the opposite side of the boulder to play the role of producers, and promptly picked up the hundred leaves, thereby receiving $100 of *income* from the sale of the food.

"What a coincidence," said Adam. "We bought $100 of output, and also received exactly $100 of income."

With patience and sensitivity, Eve made her husband see that it had to be so. "Output equals income," Eve whispered softly. "It's our first accounting identity."

They also noted that in the pre-tractor economy, total consumption was equal to total output—$100.

Then Eve said, "Let's *assume* we built a tractor this year."

"But we didn't," Adam objected.

"I'm an economist," Eve retorted. "I can assume anything I want. Now let's redo our accounting. Assume we devoted 40 percent of our work time to the construction of one tractor. Then our time spent on food production fell 40 percent, so food output fell from 100 to 60 units."

As good accountants, they reasoned as follows. Since the total labor time devoted to all production (food plus tractor) was the same, total output should still be valued at $100. Since output of the consumer good fell from $100 to $60, the price of the tractor should be set at $40, so that total output would remain $100.

"Let's call food the 'consumer' good," expounded Eve, "because it is used up—consumed—in the year it is produced. Let's call the tractor the *investment* good, because it raises productive power in future years.

"I have another accounting identity," she then exclaimed. "Output ($100) equals consumption ($60) plus investment ($40)."

Then they performed their ritual at the boulder under the tree. Two goods were now available for sale: the consumer good (food), and the investment good (the tractor). Playing the role of consumers, Adam and Eve paid $60 for the food. Playing the role of investors, they paid $40 for the tractor. Immediately, they circled to the opposite side of the boulder and, as producers, promptly received $100 of income.

"Notice," said Eve, "that we consumed $40 less than our income; our income was $100, and our consumption was only $60. I propose that we *define* income minus consumption to be *saving*. Our saving, therefore, was $40."

"What a coincidence," said Adam. "Our investment was also $40."

Once again, with gentle patience, Eve tried to explain to her husband that it had to be so, that saving had to equal investment. But this time she found it harder to make him grasp the point.

Suddenly she exclaimed, "A bank! I'll show him with a bank."

Eve promptly declared a nearby boulder to be a bank. She said to Adam, "Let's do the ritual again."

But this time when, as producers, they received the $100 of income, Eve said to her husband, "Let's save $40 of our income. Let's put $40 in that bank over there."

She led Adam by the hand to the bank and deposited the $40 on top of the bank boulder. "We've just saved $40 of our income. We'll consume the remaining $60.

"Now," she said, "let's play the role of business managers who want to buy the tractor, thereby investing $40 in new machinery. We'll borrow $40 from the bank." At the bank boulder, she took the $40 that had been saved and, leading Adam back to the original boulder, used the $40 to buy the tractor.

"Investment equals saving," she whispered softly. "It's our third accounting identity."

Then Eve sat under the tree of economic knowledge, lost in thought. Suddenly, she jumped to her feet.

"Do you realize what this means, dear husband? Investment, unlike consumption, raises our future productive power. But to invest, we must save. And saving means consuming less than our income."

"And there isn't any way around it?" asked Adam once again.

"None whatsoever," replied Eve. "You don't get something for nothing," she added, her cheerful tone confirming that she was, indeed, the first economist.

Adam grumbled, "You certainly do practice a dismal science."

Optimal Saving in a Lonely Eden

Adam paced anxiously under the tree of economic knowledge.

"What's the matter, dear?" asked Eve.

Adam's eyes darted in all directions.

"I don't think I want to build the tractor," he confessed guiltily. "Of course, I want the tractor. But I don't want to save for the tractor. I can't bear to cut my consumption," he blurted out, tears streaming down his face.

"There, there, dear," Eve comforted him. "I know how much trips to our little food mall mean to you. You've become quite attached to it."

"I'm addicted to it," Adam cried out in despair. "I don't think I can survive a cut in my consumption, even for one year. Eve, I need a compassionate therapist, and curse my lot, the only other living being is an economist."

Without taking offense, Eve replied, "Now, dear, I know what you've heard about economists—that they are devoid of emotion, that they depict man as a calculating robot, that all they care about are dollars and cents. But you forget that all this came later, after Adam Smith glorified the virtues of specialization. For God's sake, this is Eden, we're the only two people on earth—at least, as far as we know—so I can't afford to be only an economist. I also have a degree in psychology. Let me help. Here, lie down on this couch, and try to relax."

Eve's tender, soothing voice convinced Adam that she was, indeed, more than just an economist. Maybe she could help him with his overwhelming guilt. Adam lay down, without even asking where the couch came from.

"Eve," Adam confessed, "I know that economics teaches that saving is always better than consuming. I know my lust for consumption is wrong. But I can't repress it."

"Adam, this may come as a pleasant shock to you, but that is not what modern economics teaches. Saving is not always better than consuming."

Adam was indeed shocked. "I don't understand," he gasped with relief.

"Believe it or not," Eve continued, "economics cannot tell you whether to save or not. Economics simply shows you the consequence. It's then up to you to decide."

"What do you mean by 'consequence'?" Adam asked.

"Well," answered Eve, "here's how an economist would help you make your decision. First, she would ask: Once the tractor is built, how much will it raise food output in each future year?"

"In my dream," Adam replied, "the tractor raised food output $20 per year, forever."

"How convenient that God has not yet cursed mankind with depreciation," Eve laughed. "Since the tractor lasts forever—it never depreciates or wears out—we can easily compute the *rate of return* on our saving. The tractor will cost us $40 of consumption, but it will then yield a return of $20 per year, so the rate of return is $20 divided by $40, or 50 percent."

Agitated, Adam asked, "Doesn't economics teach that you should save if the rate of return is 50 percent?"

"No," Eve replied in a soothing voice. "Economics simply asks you to compare, in your mind, two situations. Under the first, without the tractor, you consume $100 of food in every year. Under the second, with the tractor, you consume $60 in the first year, and then $120 in all future years. Economics then says that there is no right or wrong choice. Whatever you decide is okay with economics. Just as economics does not presume to tell you whether to eat fewer apples and more oranges, so economics does not presume to tell you whether to consume less in the present and more in the future."

"You mean, economics accepts *consumer sovereignty* and applies the principle to present versus future consumption, as well as to apples versus oranges?" Adam asked with a feeling of relief.

"Yes," Eve laughed, gently wiping the perspiration off her husband's forehead with a fig leaf. "You can check any standard economics text. In fact, I'll draw you another production possibilities curve to you show the trade-off we face."

With a stick Eve drew Figure 1.2 in the dirt. She explained, "As in Figure 1.1, the horizontal axis shows the amount of food produced and consumed this year. But this time, the vertical axis shows the amount of food produced and consumed next year. If the amount of food is 100 this year, because no tractor is built, then the amount of food will also be 100 next year. I label this point A, corresponding to point A in Figure 1.1. But if the amount of food is cut to 60 this year, enabling the construction of 1 tractor, then the amount of food next year will be 120. I label this point B. It corresponds to point B in Figure 1.1."

"I can't bear to think of it," said Adam, "but what would happen if we build two tractors?"

"We would go to point C. This year the amount of food would be only 20."

"I can't look at point C," said Adam trembling, shutting his eyes.

"It's okay," Eve reassured him, "I promise you we won't go to point C, or even to point B if it upsets you so much. If you want to stay at point A, then that's where we'll stay."

"I do, I do," cried Adam. "But tell me, if we did go to point C by cutting our consumption to 20 this year and building two tractors, how much would we be able to consume next year?"

"You might think it would be 140," said Eve, "but I estimate it would be 136, so that's what I indicated in Figure 1.2."

"But why only 136?" asked Adam. The first tractor raised the amount of food by 20, from 100 to 120. So wouldn't the second tractor raise the amount of food by 20 again, from 120 to 140?"

"Probably not," replied Eve. "The second tractor would plough lower quality land. And, no offense, my dear, but you would drive the second tractor, and, well, let me put this as gently as possible, I'm afraid you don't drive as well as I do."

"In other words," said Adam, "the second tractor would raise food

Figure 1.2 **Another Production Possibilities Curve**

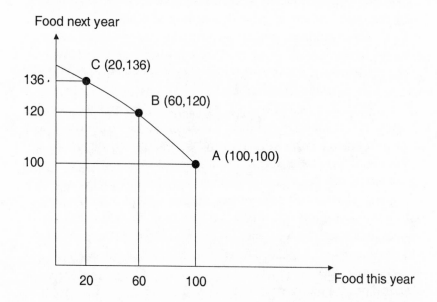

Food next year

C (20,136)

136

B (60,120)

120

A (100,100)

100

20   60   100

Food this year

output by only 16, not 20, because it would be working with lower quality land and labor."

"I'm afraid so," said Eve. "I'm going to call this phenomenon *diminishing returns.* I'm labeling point C in the diagram (20,136). The rate of return on the investment in the first tractor would be 50 percent (20/40), but the rate of return on the investment in the second tractor would be only 40 percent (16/40)."

"I'm going to peek at point C," said Adam nervously. Slowly he opened his eyes. "Eve, you've connected point A (100,100), point B (60,120), and point C (20,136), and this time it really is a curve, not a straight line."

"That's right," said Eve. Sometimes a production possibilities curve is a straight line, as in Figure 1.1, but sometimes it's a curve, as in Figure 1.2."

Adam shut his eyes again, composed himself, and slowly asked Eve, "So I can decide whatever I want, as long as I recognize the trade-off, and economics will not condemn me?"

"Yes, my love," Eve replied.

"I can decide to stay at point A, instead of moving to point B, or, God forbid, point C?"

"Yes, my love," Eve said softly. "We'll stay at point A, and economics will not condemn us."

Far East of Eden

And so Adam and Eve lived happily, as though the dream had never occurred, until one day, while wandering far east of Eden, Adam discovered a large island shaped like the letter *J*. One of Adam's greatest pleasures was naming things and places, so when he climbed the island's highest mountain, he exclaimed, "Let this be Mount Fuji, and let this island be J-Land." Exhilarated by the view from Fuji's summit, Adam gazed down into the valley.

"I am truly happy," he said to himself.

But at that very moment, Adam received a shock. Was that smoke? Could it be? Adam squinted. Were those two human forms? Adam pulled out his pocket Bible and reread the first chapter. Then he squinted again and stealthily descended the mountain to get a better look. Behind a huge rock, Adam trembled. Less than a hundred yards away was another human couple.

With fear, Adam whispered, "Let them be couple J."

Adam eavesdropped, and what he heard struck terror into his heart. Couple J had somehow obtained tractor instructions and was contemplating the same decision: to save or not to save. To his horror, Adam overheard the decision: couple J would save enough for not one but two tractors. When darkness fell, Adam fled the island and raced to Eden to report the news. An agitated discussion ensued.

"We should ignore what the Js do," said Eve firmly. "We made our decision. Let them make theirs. If they are willing to make the present sacrifice, let them enjoy a higher standard of living in the future."

As the first economist, Eve pointed out that her subjective "utility" (satisfaction) should depend only on her own (and Adam's) consumption, not on the Js' consumption. In fact, she planned to write a standard economics textbook postulating that a person's utility depends only on his own consumption, not on anyone else's.

"Relativity has no place in standard economics," she insisted.

But Adam was not so sure. He said, "This may be a bad time to

bring it up, my love, but do we plan to have children? Because if we do, we're going to have a problem. Our children are bound to learn about couple J's children. After all, communication and transportation are bound to improve, and the Js may even come to Eden as tourists. What will we tell our children when they discover that they consume less than J children because we sacrificed less than J parents?"

"We'll tell them," Eve replied adamantly, "that they should avoid standard of living comparisons. We'll teach them that envy is wrong. If all else fails, we'll show them that standard economics textbooks postulate that individuals are unaffected by relativity."

But even as she spoke these words, Eve began to feel uneasy. She was a good economist who had mastered the standard framework, but she retained an open mind about its assumptions.

"What if relativity does matter to our children?" she asked herself. "What if human nature cannot be purged of relativity?"

Adam continued, "There is a certain irony in the confession I'm about to make, but I've been reading some works in evolutionary biology. The argument goes like this. Suppose an individual who monitors his relative position and adjusts his effort accordingly— intensifying it when he's falling behind, relaxing it when he's ahead— is more likely to survive when a crisis, such as famine or predator attack, occurs. If so, then a relativity emotion may be selected for by, dare I say it, a Darwinian process. We may be unable to purge our descendants of relativity. If their standard of living is worse than the Js', they will be unable to ignore it, no matter what we say."

Now it was Eve's turn to be glum. "Maybe we'd better match the Js' saving," she said.

Eve's Dream

That night it was Eve's turn to dream. In her dream, she envisioned herself on Mount Fuji, watching the Js below. How they sacrificed in the year the tractors were built! They cut their consumption to $20, saved $80, and built two tractors worth $40 each. But then Eve watched them, the next year, use the two tractors to plant, plow, and harvest. The Js and their children now enjoyed much more than $100 of consumption, and would do so forever. And the additional con-

sumption was not simply food. Somehow, the Js were making and consuming dazzling appliances, undreamed of in Eden.

"No!" Eve cried in her dream. To Eve's horror, she saw the Js board an airplane and travel to Eden. With them they brought their appliances, not for sale, but for show. The children of Eden were filled with awe and wonder. "Why don't we have those?" they asked. Eve saw an expression of sorrow come over the Js' faces. They pitied the poor children of Eden.

"No!" Eve screamed, and she awoke, drenched in perspiration. "We can't let it happen. We can't let the Js move ahead of us," she cried. The morning sun had just risen over the horizon.

"Everything has changed," Adam said grimly. "Woe, that I ever set eyes on the J island."

APPENDIX

There's Just No Substitute for Our Own Saving

Resolving to match the Js, Adam and Eve fell into a deep sleep. In the morning, Adam awoke and suddenly exclaimed with joy, "We don't have to save! We don't have to save!"

"What do you mean?" asked Eve sleepily.

"We can borrow from the Js to finance a tractor! If we borrow, we can have our tractor, and we won't have to cut our consumption below $100 in the year we acquire it!"

"So," said Eve, "you think that our solution is simply to borrow from J-Land?"

"Well, isn't it?" asked Adam, suddenly growing worried.

"Poor dear," said Eve. "Adam, my love, sit down and have some breakfast. This is one lecture you don't want to hear on an empty stomach."

Adam ate nervously, and then Eve began.

"Suppose that the quantity and quality of land and labor are the same in Eden and J-Land. But suppose we save nothing this year, and the Js save for two tractors. If our two economies were isolated, then beginning next year, obviously two tractors would operate in J-Land, and none in Eden, so J-Land's output would be greater than Eden's.

"But," continued Eve, "suppose our two economies interact. Then I will show you, in a moment, why one of the two tractors would be invested in Eden. In fact, in future years, output in J-Land and Eden would be identical, because both economies would have one tractor."

"Yes!" shouted Adam. "That's what I was hoping!"

"Adam, my love, don't celebrate too soon. Let me continue. Remember we said that the rate of return on one tractor would be 50 percent, because it costs $40, but raises output $20 per year. Well, that's true whether it operates in J-Land or Eden. But if a second tractor is used in J-Land, then the return on the second will be less— for example, 40 percent."

"Why?" asked Adam.

"The reason, Adam, is simple. The first tractor will work the best land in J-Land. If a second tractor also operates in J-Land, it will work land that is not quite as fertile, so its return will be less. I call this phenomenon *diminishing returns*—remind me to emphasize it when I write my economics textbook.

"But," Eve continued, "if the second tractor is used in Eden, we'll use it to work our best land, and its return will be 50 percent. So if the second tractor is invested in Eden, instead of J-Land, it will yield a higher return—50 percent instead of 40 percent. Now, Adam, suppose for a moment you were the Js, and you were willing to save $80. Would you use your $80 to finance two tractors in J-Land? Or would you use your $80 to finance one tractor in J-Land and one in Eden?"

"Why, the answer is obvious even to me," exclaimed Adam. "I'd split my $80 between one tractor in each country, because then each $40 would earn a 50 percent return. If I put all $80 into tractors in J-Land, the first $40 would earn a 50 percent return, but the second $40 would only earn a 40 percent return."

"I'm proud of you," said Eve. "To obtain the higher return, the Js will lend us $40 to obtain one tractor for use in Eden. Thus, in future years, both economies will operate with one tractor, and use it to work its best land, so our output and J-Land's will be identical."

"But," Adam said, "I'm still shaky on the mechanics of how saving in J-Land finances investment in Eden."

"Of course," replied Eve patiently. "Let's elaborate our ritual."

She led Adam to the two boulders and reminded him that one was a bank.

"This bank," Eve explained, "would receive all saving, and lend it to investors to purchase tractors. In the year when saving and investment occur, both economies produce $100 of output and income. We save nothing, and the Js save $80—enough to finance two tractors, $40 each. The Js bring $80 of saving to the bank. The bank lends it out to whoever can use it to generate the highest return."

"Why does the bank care about who can generate the highest return?" asked Adam.

"Because, my love," Eve replied, "the bank will charge interest on the loan to the borrowers, and the higher the return on the investment, the higher the interest the bank can collect. The J savers, of course, will ultimately receive this interest.

"Now," Eve continued, "imagine that you and I go to the bank as investors seeking to borrow $40 to buy a tractor, and J investors with the same motive come to the bank as well. On the first tractor, we and the Js will each generate a 50 percent return. Then the bank can charge us interest of just under 50 percent, and we will still find it worth borrowing, because the return on the tractor will slightly exceed the interest we must pay the bank. But J investors who seek a second tractor will find the interest rate too high, because they can generate only a 40 percent return on the second tractor. So we will get the loan, not them."

"I feel sorry for the Js," Adam laughed. "We save nothing, they save $80 for two tractors, and yet one tractor gets invested and used in both economies, so that we can produce the same output as they can in future years. Thrift is folly."

"Ah, Adam, I'm afraid you've missed the crucial point. Let's go a step further. Just as the tractor is 'permanent' and never wears out (recall that God has not yet cursed mankind with depreciation), so that a $40 tractor raises output by $20 in every future year, let's assume that the bank is willing to give us a permanent loan so that we never have to repay the principal—the $40—but must pay a constant amount of interest in every future year. Having permanently borrowed $40, in each future year we will owe the bank nearly $20 per year of interest, because the bank's interest rate is just under 50 percent. Who do you think will ultimately receive this interest? The Js, who saved in the initial year. In all future years, interest payments will flow from Eden to J-Land through the bank.

"Let's suppose," Eve continued, "that there is no further saving by anyone in future years. Both economies will produce the same output—$120 per year of food—because the $40 tractor has raised output from $100 to $120 in both economies. But although one tractor is used in each economy, in effect the Js own both tractors. Because of their saving in the initial year, they receive the return from both tractors. Thus, we will continue to consume only $100, because we must pay $20 of interest to the Js, through the bank, while the Js will consume $140 per year—the $120 they produce in J-Land, plus the $20 they can buy with our interest payments."

At last Adam seemed to follow Eve's analysis. "If someone looked at our two economies in future years," he said, "they might think we have the same standard of living because we each use one tractor, so our machinery per worker is the same, and our output per worker—our labor productivity—is the same. But they would be mistaken."

"Exactly right," said Eve. "What counts is who owns the capital equipment—the tractors. The Js own both tractors even though each economy uses one in production. Thus, the Js have a *wealth* of $80, and we have no wealth. As a result, the Js' wealth generates $40 of capital income, a 50 percent return on their wealth. Both we and the Js earn $100 of labor income, but the Js earn $40 of capital income. Thus, the Js' total income is $140 while ours is $100, and in future years, their consumption will be $140, while ours will be $100."

Eve continued. "Wealth is *net worth*. We possess $40 of assets—one tractor—but our liabilities are also $40 because we borrowed $40 from J-Land; so our net worth (assets minus liabilities) is zero. By contrast, the Js' net worth is $80. They possess a $40 tractor, and they are also owed $40 by us. In fact, we will give them a $40 Eden bond to hold to indicate this, so their total assets are $80. With zero liabilities, their net worth is $80. Hence, our wealth is zero, and theirs is $80. And the difference in wealth is, of course, due to the difference in saving."

"So," continued Adam, "we would be in debt to J-Land. We would be a debtor nation, because our wealth—zero—would be less than our capital stock—$40. And J-Land would be a creditor nation, because its wealth—$80—would exceed its own capital stock—$40. So every year, debtor nation Eden would make interest payments of $20 to creditor nation J-Land."

"That's right," replied Eve.

"Then," concluded Adam, "what matters for our future standard of living is the saving we do, not the investment that occurs within our borders."

"I'm proud of you," exclaimed Eve. "Suppose we tried to ignore the fact that the Js saved $80 and we saved nothing. We might try to defend ourselves by pointing out that investment was $40 in both economies. Each country obtained one tractor. Then, in future years, we could point out that both economies produced $120 of output. But, in fact, their standard of living would be higher, because they would have more wealth and would receive interest payments from us, and therefore could afford more consumption."

"I have one last question," said Adam with new confidence. "One $40 tractor raises output $20 per year—a 50 percent return. You assumed that the J savers will capture the full return, so we will pay them a 50 percent interest rate, or $20 per year. But isn't it possible they will capture only a fraction of the 50 percent return, and we will capture a fraction ourselves? For example, maybe we will only pay them 25 percent (not 50 percent) of $40, or $10 per year, and keep $10 per year for ourselves."

"I have a confession," said Eve. "You are right. If the tractor raises the *marginal* product of our labor or land (the increase in output due to the last unit of labor or land), then we will capture a fraction of the return, so the outcome won't be so bad. I'll explain why, any day you are willing to hear a lecture on the marginal productivity theory of income distribution."

"But," continued Eve, "my simplification is justified because it doesn't change the basic point. If they save and we don't, then even if the Js capture only a fraction of the full return, their future standard of living will still exceed ours. The reason is simple. The Js, who own all labor, land, and capital used in J-Land, will obviously capture the full return on the $40 tractor invested in J-Land. And as long as J savers capture some of the return on the $40 tractor invested in Eden, we will capture only a fraction of the full return. So due to the two tractors, the Js' future standard of living must be higher than ours, even if we capture a fraction of the return on our $40 tractor."

Adam looked glum. Finally, he said, "Eve, I understand that if the Js save and we don't, they'll live better than we will because we'll

pay them interest every year. But exactly how would they get more of our crop to consume? What would happen to exports and imports?"

"Let's trace it through," said Eve. "In the year of saving and investment, assume one tractor is built in each economy, so each economy produces $60 of food, and a $40 tractor. But we borrow $40 from J savers through the bank. We use the $40 to keep our consumption at $100, while the Js reduce their consumption to $20, saving $80 of their income. Clearly, we must import $40 of food from J-Land to keep our consumption of food at $100, and the Js would export $40 of food and consume only $20 of food."

"So," said Adam, "in that first year we would run a trade deficit of $40, and J-Land would run a trade surplus of $40."

"But," said Eve, "let's now consider each future year. Both economies produce $120 of food, but we consume $100, and the Js consume $140. Clearly, we must export $20 of food to J-Land. So in each future year, we would run a trade surplus of $20, and J-Land would run a trade deficit of $20. Yet they would have the higher standard of living."

"But the trade balance leaves out something important," exclaimed Adam. "It includes only payments from the flow of goods—$20 of food. But while we're receiving $20 from the Js for food, we're paying $20 to the Js for interest on the original $40 loan. Every year, the Js use the $20 of interest they receive from us to buy $20 of food from us."

"Adam, that's impressive reasoning," smiled Eve. "You're right. We do need a balance that includes interest payments as well as payments for goods. I propose we call it the *current account* balance. In each future year, our current account balance would be zero, because our receipts from $20 of food would be offset by our $20 of interest payments."

"But," sighed Adam, "even though our current account balance would be zero, and our trade balance would be a surplus of $20, our standard of living would be lower than J-Land's."

"Exactly right," said Eve with a smile.

"So," said Adam, "it would be a mistake for us to feel good about our current account balance of zero and our trade surplus of $20, because what really matters is that the Js would be enjoying a higher standard of living, since they had saved and accumulated more wealth.

So even if the headlines focus on the trade and current account balances, the most important thing to watch is who is saving more and accumulating more wealth, because that's what clearly tells who will enjoy the higher standard of living in the future."

"Adam," said Eve with obvious affection and tears in her eyes, "you've made me truly happy. You've grasped some basic economics."

Adam and Eve went outside, and stood silently looking at the far eastern horizon where they could see the snowy peak of Mount Fuji faintly in the distance.

"I pledge to you, Eve, that we will save enough to finance not one, but two tractors. We will match the Js in wealth accumulation. Our children's standard of living will be second to none in the world."

Together, Adam and Eve walked bravely toward the sunrise.

II | Microeconomics

2

Markets: Demand and Supply

How does a *competitive market* work? A competitive market, which consists of many sellers competing to sell a specific product to many buyers, is governed by *demand and supply*.

Demand and Supply

Let's illustrate with the market for gasoline, shown in Figure 2.1. A "market" consists of buyers (demanders) and sellers (suppliers).

The lower the price, the greater is the quantity demanded by the buyers of gasoline (incidentally, the word "demand" does not imply rudeness). In Figure 2.1, if the price is $2.00 per gallon, consumers want to buy 80 gallons; but if the price is $1.50, they want to buy 100 gallons, and if the price is $1.00, they want to buy 120 gallons. Connecting these three points results in the *demand (D) curve*, as shown.

The higher the price, the greater is the quantity supplied by the sellers of gasoline (I'll explain why shortly). In Figure 2.1, if the price is $1.00, the producers want to supply 80 gallons; but if the price is $1.50, they want to supply 100 gallons, and if the price is $2.00, they want to supply 120 gallons. Connecting these three points results in the *supply (S) curve*, as shown.

Thus, Figure 2.1 shows the demand for, and the supply of, gasoline. Economists label the demand curve D and the supply curve S,

Figure 2.1 **Demand and Supply**

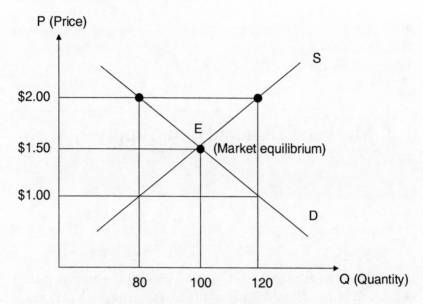

and call them "curves" because they don't have to be straight, even though they are in this example.

So what will the price of gasoline turn out to be in this market? The answer is $1.50 a gallon, the price at which the demand curve and the supply curve intersect. Why?

Suppose the sellers of gasoline initially set the price at $1.00. At that price, the buyers demand 120 gallons, but the sellers are only willing to supply 80 gallons. It doesn't take long for the sellers to see that at $1.00, there are desperate buyers who can't get as much gasoline as they want. With *excess demand* for gasoline (120 − 80 = 40), sellers realize they can get away with raising the price. So they do. As the price rises, sellers are willing to supply more gasoline, and buyers reduce the amount they demand. When sellers set the price at $1.50, they are willing to supply exactly as much as the buyers demand—100 gallons. The sellers notice that there are no longer any desperate buyers. So they leave the price at $1.50.

Symmetrically, suppose the sellers of gasoline initially set the price at $2.00. At that price, the buyers demand only 80 gallons, but the sellers want to supply 120 gallons. Now some sellers are

desperate. With *excess supply* of gasoline (120 − 80 = 40), sellers realize they must cut the price to sell as much as they want. So they do. As the price falls, the sellers reduce the amount of gasoline they are willing to supply, and buyers increase the amount they demand. When sellers set the price at $1.50, they are willing to supply exactly as much as the buyers demand—100 gallons. So they leave the price at $1.50.

Economists call the point where D and S intersect the *market equilibrium*, and in Figure 2.1, I have labeled it point E. The market will go to point E: the price will eventually settle at $1.50, and the quantity actually bought and sold will settle at 100. Rather than use the word "equilibrium," I will simply say, "the market moves to point E where D and S intersect."

Now let's explain why the higher the price, the greater is the quantity that sellers want to supply. Sellers compare the cost of producing another gallon—the *marginal cost*—to the price they will get for it. As long as the marginal cost is less than the price, they make more *profit* by producing another gallon. Profit equals total sales revenue minus the total cost of production. Profit is what the sellers get to keep after paying their costs. For example, when the price is $1.00, sellers keep producing as long as the cost of an additional gallon is less than $1.00; according to the diagram, 80 gallons. But there are some additional gallons that cost between $1.00 and $1.50 to produce. If the price rises to $1.50, it will now be profitable to produce and supply these additional gallons; according to the diagram, 20 additional gallons (for a total of 100). Similarly, there are some additional gallons that cost between $1.50 and $2.00 to produce. If the price rises to $2.00, it will now be profitable to produce and supply these additional gallons; according to the diagram, 20 additional gallons (for a total of 120). Hence, the higher the price, the greater the quantity the sellers find it profitable to supply.

Shifts of Demand and Supply

What causes the market price to change? Anything that shifts either the demand curve or the supply curve.

The demand curve shifts to the right if something makes the buyers demand more at the same price. What might do this? One thing is

Figure 2.2 **Shift in Demand**

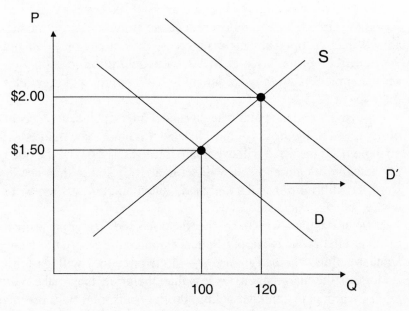

an increase in the income of the buyers. Another is an increase in preference for the good. So if the income of the buyers rises, or people prefer to drive more, the demand curve shifts to the right, as shown in Figure 2.2, from D to D'. The result is a rise in the market price from $1.50 to $2.00 and a rise in the quantity actually purchased from 100 to 120 gallons.

The supply curve shifts to the right if something increases the total amount supplied at the same price. What might do this? Suppose that following the rise in price from $1.50 in Figure 2.1 to $2.00 in Figure 2.2, new business firms enter the industry, attracted by the higher price and profits. Each old firm produces the same amount, but with new firms, the total amount supplied increases at any given price so, as shown in Figure 2.3, the supply curve shifts from S to S'. The result is a fall in the market price from $2.00 to $1.50 and a rise in the quantity actually purchased from 120 to 140.

Now go back to Figure 2.1 with the market at point E, so the price is $1.50 and the quantity 100. Assume the D curve does not shift. What happens if the cost of production falls, for example, because a technological advance reduces the cost of getting oil out of the ground

Figure 2.3 **Shift in Supply**

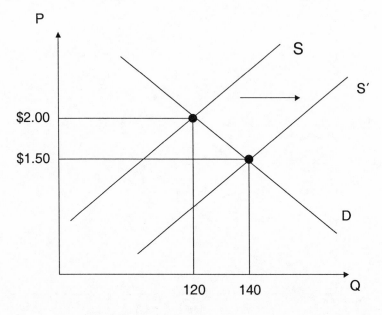

or refining oil into gasoline? Then the supply curve shifts down, as shown in Figure 2.4. Why? Before the technological advance, the cost of producing the 100th gallon was just under $1.50, so at a price of $1.50, the sellers found it profitable to supply 100 gallons. With the technological advance, suppose the cost of producing each gallon falls $0.50—for example, the cost of producing the 100th gallon falls from $1.50 to $1.00. Hence sellers would be willing to supply 100 gallons if the price were $1.00. Thus, the $0.50 reduction in cost causes the supply curve to shift down $0.50. *Note that the height of the supply curve equals the marginal cost*—the cost of producing the last gallon. If a technological advance reduces marginal cost by $0.50, it will shift down the supply curve $0.50, as shown in Figure 2.4, from S to S'. The result is a fall in the market price from $1.50 to $1.25 (note that the price falls less than $0.50, as you can see in the diagram—exactly how much it falls depends on the slopes of the D and S curves—the fall of $0.25 is just for illustration) and an increase in the quantity actually purchased from 100 to 110.

Symmetrically, starting from point E in Figure 2.1, what happens if the cost to the sellers rises, for example, because the government lev-

Figure 2.4   **Cost Reduction**

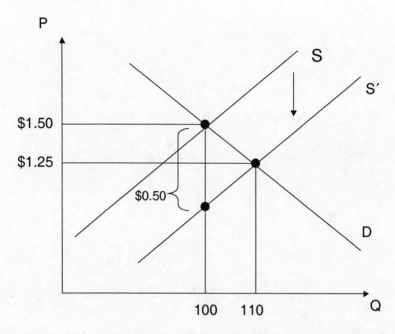

ies a tax per gallon on the sellers. Then the supply curve shifts up, as shown in Figure 2.5. Why? In Figure 2.1, before the tax, the cost of producing the 100th gallon was just under $1.50, so at a price of $1.50, the sellers found it profitable to supply 100 gallons. Suppose the tax is $0.50 per gallon. Then the cost to the sellers of supplying the 100th gallon rises from $1.50 to $2.00 because the sellers must pay the government $0.50 a gallon. Hence sellers would be willing to supply 100 gallons if the price were $2.00. Thus, the $0.50 increase in marginal cost causes the supply curve to shift up $0.50, as shown in Figure 2.5, from S to S'. The result is a rise in the market price from $1.50 to $1.75. Note that the price rises less than $0.50, as you can see in the diagram—exactly how much it rises depends on the slopes of the D and S curves—the rise of $0.25 is just for illustration. Also for illustration, the quantity actually purchased decreases from 100 to 90.

Figure 2.5 is a good place to see the effect of the *elasticity* of demand. Economists say demand is elastic if a rise in price causes buyers to sharply reduce the quantity they want to buy—in other words, demand is elastic if the demand curve is flatter than shown in

Figure 2.5 **Impact of a Tax**

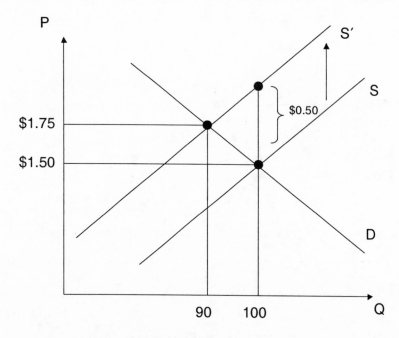

Figure 2.5; if the D curve were flatter, then in response to the $0.50 tax, the price would rise less than $0.25. Symmetrically, economists say demand is inelastic if a rise in price causes buyers to slightly reduce the quantity they want to buy—in other words, demand is inelastic if the demand curve is steeper than shown in Figure 2.5; if the D curve were steeper, then in response to the $0.50 tax, the price would rise more than $0.25.

Free Market Versus Price Control

Consider the market for rental apartments near a university. Initially, the market is at point 0 in Figure 2.6, the intersection of D and S; the monthly rent is $400 and there are 1,000 apartments rented. Suppose the university suddenly decides to expand its programs and student enrollment. The demand curve shifts to the right from D to D' and landlords raise the rent to $600, so the market moves from point 0 to point 1. In the *short run*, supply is *inelastic*—that is, even a significant rise in price causes only a small increase in supply; in

Figure 2.6   **Free Market Versus Price Control**

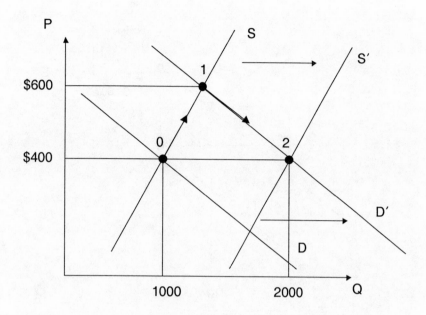

Figure 2.6, the supply curve is very steep, because it takes a long time to construct new apartments. In the short run, the high rent induces some homeowners to rent out a room to a student, and this accounts for the small increase in supply. In the short run, landlords make large profits, and students are unhappy about high rents and crowded living conditions (many squeeze into old apartments and old university dorms).

But in the *long run*, the high rent and large profits attract *new entrants* into the market—apartment builders who construct new apartments. Gradually, the supply curve shifts to the right, as shown in Figure 2.6, from S to S', the rent comes down, and the market moves gradually down the D' curve from point 1 to point 2. The construction boom continues until the rent comes down enough to return profits to normal. Thus, it is possible that in the long run the rent may come back down to $400, as shown, if the market settles at point 2.

With a free market, then, in the short run consumers are unhappy, facing high rents and crowded conditions, but in the long run they

are pleased, because there are many more apartments (2,000 instead of 1,000), living conditions are no longer crowded, and rent is back to normal.

Suppose, however, that in the short run unhappy students persuade the city council to pass a rent control law prohibiting any landlord from charging more than $400. Then with this price control, price and quantity stay permanently at point 0. Without higher rents and large profits, there are no new entrants into the market—there is no construction boom. The supply of apartments remains 1,000. Year after year, many students are not able to find apartments they want (demand), many live in crowded conditions and complain about a permanent shortage of apartments.

There's an important lesson to be learned. In the short run, it is tempting for consumers to lobby for a rent control law. But in the long run, they are better off if the market is allowed to work freely without government price control. In the short run, landlords make large profits, and this may seem unfair to students paying high rents and living in crowded conditions. But the large short-run profits are a necessary part of the solution, because it is the large profits that attract new entrants and induce the construction boom. Eventually, profits and rents return to normal, and students enjoy a greater supply of apartments. For this reason, virtually all economists generally oppose price controls and favor letting the market work freely.

But what if the free market sets some prices so high that poor people can't afford necessities? Most economists agree that if we want help poor people, we should do it by getting them enough income so they can afford necessities. How? By having the government raise tax revenue and transferring cash to them. Economists, like other citizens, disagree about how much taxing and transferring should be done. We'll discuss this in chapter 9. But most economists agree that this method is generally better than price control.

Does the Market Generate the Right Amount of Each Product?

Does the market generate the right ("socially optimal") amount of good X—not too much and not too little? It usually does. Why? To understand why, we need to think about the height of the demand

curve and the height of the supply curve. Look back at Figure 2.1 for the discussion that follows.

We saw, above, that the height of the supply curve equals the *marginal cost* (MC). For example, in Figure 2.1, at the 80th gallon, the height of the supply curve is $1.00 because the cost of producing the 80th gallon is $1.00; at the 100th gallon, the MC is $1.50; at the 120th gallon, the MC is $2.00. In this chapter, we assume that the production or use of gasoline does not generate any environmental harm and that there is no tax on gasoline, so that the marginal social cost (MSC) is the same as the marginal private cost (MPC) faced by the business firm. In chapter 3 on pollution, we alter these assumptions.

The height of the demand curve equals the *marginal benefit* (MB) to drivers of each gallon—the maximum amount that they would be willing to pay for that gallon. Why? Consider the 80th gallon in Figure 2.1. If the price were $2.00, drivers would be willing to buy 80 gallons, but not a gallon more. It must be true that the 80th gallon was a very close decision—to buy or not to buy. That means that the maximum amount drivers are willing to pay for the 80th gallon is very slightly above $2.00, or approximately $2.00. Consider the 100th gallon. If the price were $1.50, drivers would be willing to buy 100 gallons; so it must be true that the maximum amount drivers are willing to pay for the 100th gallon is roughly $1.50. Similarly, the MB of the 120th gallon must be $1.00. Thus, as the numbers of gallons drivers have already bought increases, the MB declines.

So what is the socially optimal (best) amount of gasoline? Every time another gallon is produced, labor is used that could have made something else. The MC of a gallon measures the *opportunity cost*—the dollar value of the alternative output. As long as the MB exceeds the MC, consumers gain more from another gallon of gasoline than they lose by getting less of the alternative output. Hence, 80 gallons is too little, because the MB of the 81st gallon is just under $2.00 whereas the MC is only slightly above $1.00. Symmetrically, 120 gallons is too much, because the MB of the 120th gallon is $1.00 whereas the MC is $2.00. Hence, 100 gallons is just right—the social optimum.

But this is exactly the amount the free market generates. Remember, in this chapter we are assuming gasoline involves no pollution

and there is no gasoline tax. The market goes to where D and S intersect. But interpreted vertically, the D curve is the MB curve and the S curve is the MC curve. So the market generates output to the point where, for the last unit, MB (approximately) equals MC. Note that the producers of gasoline are not trying to produce the "social optimum"; they don't compare MB with MC. The producers are trying to maximize their profit—it is the profit motive that determines the amount they want to supply at each price. Yet, as the great economist Adam Smith observed in his classic book *The Wealth of Nations* in 1776, although producers pursue only their own self-interest, as long as there is competition among them, they are led by an "Invisible Hand" to benefit consumers.

The Stock Market

The price of a share of stock in a particular corporation is set by demand and supply. The intersection of the demand curve and the supply curve determines the price. Before I can explain the stock market, I need to convey some basic information about corporations and corporate stock.

Why do corporations issue shares of stock? Suppose the managers of corporation X see an opportunity to make more profit by investing in new technology. How do they get the funds to buy the new technology? One way is to use a portion of corporation X's current profits; the corporation usually (but not always) pays out a portion of profits as *dividends* to current stockholders, pays corporate income tax, and keeps the rest in the corporation—the portion kept is called *retained earnings*, so retained earnings are one source of funds to finance the purchase of new technology. Another way is to borrow, either from a bank, or from the public by issuing bonds. The corporation is obligated to pay interest and principal on any borrowed funds. Still another way is to issue shares of stock. Anyone who buys the shares becomes a part owner of the corporation. The corporation is not required to pay owners the way it is required to pay lenders, so it is less risky for the corporation to issue stock than to issue bonds.

Once new shares of stock are sold, the corporation has its funds. But the holders of shares of stock are free to sell their shares at any time. Thus, there is a market for shares of corporate stock, and most

of the supply of shares to the market on a given day comes not from corporations issuing new stock, but from holders of "old" shares.

The main reason someone wants to buy ("demands") shares in corporation X is the hope that its price will rise; if a person sells the shares at a higher price, the person enjoys a *capital gain*. For example, if the price rises from $100 to $140, the person enjoys a capital gain of $40. Another reason to buy shares is the hope that the corporation will pay shareholders a dividend. But many buy shares in corporations that pay no dividends, so the hope of a capital gain is usually the primary motive for buying.

Symmetrically, the main reason any owner of shares in corporation X wants to sell the shares is the fear that its price will fall; if the person ends up selling the shares for a lower price, the person suffers a *capital loss*. For example, if the price falls from $100 to $60, the person suffers a capital loss of $40. Another reason to sell shares is the fear that the corporation will not pay the shareholders a dividend. But many hold shares in corporations that pay no dividends, so the fear of a capital loss is usually the primary motive for selling.

So what determines the demand for and supply of shares of stock of corporation X on a given day? Lots of things. Suppose there is an announcement today that corporation X has just invented a new product that is likely to be very profitable, enabling X to pay larger dividends in the future. This will reduce the number of holders of shares of X who want to sell—supply—shares today, so the supply curve shifts left; and it will increase the number of buyers who want to buy shares of X, so the demand curve shifts to the right. Hence, the price of a share of stock in corporation X immediately rises. Similarly, if X announces that its most recent quarterly earnings (profits) were below what had previously been expected, the S curve will shift right, and the D curve, left, so the price will fall. Or suppose the chairman of the Federal Reserve ("Fed") testifies before a congressional committee that new data show the economy has suddenly weakened, implying that most companies will suffer a fall in profits; then the price of shares in most companies may immediately fall.

Because the demand for and supply of shares of stock in any corporation are subject to hopes and fears about future prices, the D and S curves can shift suddenly by a large amount; hence, stock prices

are quite volatile, and sometimes fluctuate wildly, rising sharply one day, and plummeting the next. As Fed chairman Alan Greenspan said famously in 1996, the stock market is subject to "irrational exuberance" when the future seems bright. Unfortunately, it is also vulnerable to irrational panic when the future suddenly appears bleak.

How a Business Firm Maximizes Profit

Thus far, I have made things simple by assuming that there is a going market price, and the business manager simply decides what quantity is most profitable to supply at that price—that is, the firm is a "price taker." We saw that, to maximize profit, the firm should keep expanding the quantity (Q) it produces as long as marginal cost (MC)—the cost of the next unit—is less than price (P), and stop as soon as MC rises above P; simply put, the firm should expand Q until MC equals P.

But for most firms in the economy, there is no going market price that it must "take." The firm must be a "price maker": the higher the P it sets, the less will be the Q that buyers demand. So how does such a firm decide the P that will maximize its profit? How would you decide if you were a business manager? Think about it. Cutting your P gets you less revenue for each unit you sell, but induces buyers to buy more units of Q. But producing more Q raises your total cost. Clearly, cutting your P all the way to $0 would be a mistake. So how do you find the best P?

Table 2.1 gives you the information you need. Start at the left in the first row. If you set your P at $20, buyers would demand a Q of only 1 unit, your total revenue (TR) would be $20 (because TR equals P times Q). Since your total cost (TC) of producing a Q of 1 unit happens to be $11, your profit would be $9 (because total profit equals TR minus TC, $20 – $11). Now go to the second row for P. If you set your P at $19, buyers would demand a Q of 2 units, your TR would be $38 ($19 × 2). Since your TC of producing a Q of 2 units happens to be $24, your profit would be $14 ($38 – $24). The remaining three rows are obtained in the same way. Now look at the profit column: What's the highest number? $15. The P that gets it is $18, so that's the P you should set.

Another way to find the P that maximizes profit is to look at the two columns labeled MR and MC. MR is *marginal revenue*— the revenue from selling 1 additional unit. Imagine you set P equal

Table 2.1

Setting the Price to Maximize Profit

Buyers demand for the product				Cost of production			
P ($)	Q	TR ($)	MR ($)	Q	TC ($)	MC ($)	Profit ($)
20	1	20	18	1	11	13	9
19	2	38	16	2	24	15	14
18	3	54	14	3	39	17	15
17	4	68	12	4	56	19	12
16	5	80		5	75		5

Note: P = price; Q = quantity; TR = total revenue; MR = marginal revenue; TC = total cost; MC = marginal cost.

to $20, so buyers buy a Q of only 1 unit. If you cut your P to $19, they'd buy a Q of 2 units, and your TR would rise from $20 to $38. Hence, your MR would be $18 ($38 − $20). MC is *marginal cost—* the cost of producing 1 additional unit. If you raised the Q you produced from 1 to 2 units, TC would rise from $11 to $24, so your MC would be $13. Since MR ($18) is greater than MC ($13), you should do it—raise Q from 1 to 2 units. Should you raise Q from 2 to 3 units? Yes, because MR ($16) is greater than MC ($15). But should you raise Q from 3 to 4 units? No, because MR ($14) is less than MC ($17). Hence, a Q of 3 units maximizes your profit, so you should set P equal to $18. Here's a concise way to summarize this: Cut P and increase Q as long as MR is greater than MC, and stop when MR (approximately) equals MC (before MR falls below MC).

Dynamism of a Market Economy

A market economy is not perfect, but it is an amazing engine of progress. Pursuing self-interest, business firms compete with each other by inventing new products and new technologies, and work hard to attract consumers either with better quality or lower cost and

price. Any business firm that becomes complacent, believing it can coast on its past accomplishments, suddenly finds that new firms have entered its industry and are threatening to win consumers away. The pressure is always on, and while this can make life tough on managers and workers, it benefits consumers.

In many of the chapters that follow in this book, we will see that free markets don't solve everything, and that government intervention may be necessary to make the economy work better, or to remedy its shortcomings, or to try to make outcomes fairer. But before moving on, we should pause to recognize the dynamism of a free market economy, and realize that this engine of progress sets the stage for certain government interventions that may make our society even better. The engine should never be taken for granted, and whatever government interventions are undertaken to make things even better, care should be taken to let the dynamic engine do its work.

3

Environmental Pollution

The free market is not perfect. In fact, sometimes it fails. Perhaps the best example of market failure is environmental pollution.

A Classic Market Failure

It is sometimes claimed that economists care only about material output and not a bit about the environment. The claim is false. Virtually all economists agree, and have agreed for many years, that the free market generates too much pollution, that this is a serious failure of the market, and that something should be done about it. Now, don't get me wrong. I am not going to claim that economists, as a group, are enthusiastic environmentalists. Some are, some aren't, like most any other group of citizens. But economists agree that since many people care about environmental quality, then we must judge the economy's performance by its provision of environmental quality as well as material goods. You may not associate humility with the economics profession. But most economists do pay homage to the humble doctrine of "consumer sovereignty." This requires a word of explanation.

In standard economics, economists refuse to judge a consumer's preferences. If the typical consumer likes apples more than oranges, who are we to judge? We take consumer preferences as

given, and then ask: How well does the economy satisfy these preferences?

How do economists respond when we discover that the typical consumer likes environmental quality as well as apples, oranges, and other material goods? Naturally, we wonder who we are to judge. So we give the preference for environmental quality the same respect as the preference for material goods.

Then, unanimously, we give the free market a failing grade for the poor environmental quality it generates. Even the most ardent promarket economists give it a failing grade in this department. But then economists go a step further. We locate the source of the "market failure." The problem is that no one owns the air and water. There is a failure of property rights. Let me explain.

Whenever something is "free," it is used wastefully. But anything that is owned is seldom free. Naturally, the owner insists on charging a price for its use, so any potential user is deterred from frivolous use. But who owns the air above city X, or the water in river Y? No one owns it, so no one charges a price for "using" it—that is, polluting it. Is it any wonder, then, that it is polluted excessively under a free market?

In chapter 2, we explained why the free market generally produces the right amount of each product—not too much and not too little. But this is true only if producers must pay the full *marginal social cost* (MSC). Consider the chapter 2 example—gasoline. We now assume that gasoline use generates pollution and that the environmental harm per gallon is $1.00. We will call the amount of pollution generated by using a gallon of gasoline *one unit of pollution*, so each gallon of gasoline generates a unit of gasoline pollution. Note two points about this assumption. First, to simplify our example, we ignore the fact that different kinds of gasoline generate different amounts of pollution per gallon; we assume there is just one kind of gasoline and its environmental harm per gallon is always $1.00. Later we'll discuss what should be done when the amount of pollution per gallon varies according to the kind of gasoline. Second, what does it mean to say the harm per gallon is $1.00? This is a difficult question to answer. Again to simplify, one answer might be: Suppose health experts tell people the health consequences per unit of gasoline pollution, and then ask people how much they would be willing to pay

to reduce gasoline pollution by one unit; and suppose that people reply, $1.00. Then we might say that the harm per gallon is $1.00.

If the environmental harm (cost) per gallon is $1.00, then, as shown in Figure 3.1, the marginal social cost is $1.00 greater than the *marginal private cost* (MPC) actually paid by producers (sellers) of gasoline. What producers supply, of course, depends only on the MPC—the height of the supply curve equals the MPC they pay. The free market goes to the intersection of D and S, and without any government environmental policy S coincides with MPC, not MSC. So too much gasoline (100 gallons) is sold and used. The best ("socially optimal") amount of gasoline is where the MB (marginal benefit) equals MSC— where the MB (D) curve intersects the MSC curve—80 gallons in Figure 3.1. The free market generates 20 gallons more than the socially optimal amount (100 instead of 80).

The shaded triangle shows the dollar value of the harm to consumers of these 20 extra gallons. Why? Consider the 100th gallon. The height of the D curve, $1.50, equals marginal benefit (MB), the maximum amount consumers would pay for the 100th gallon. But the marginal social cost of the 100th gallon equals $2.50—it is $1.00 greater due to the environmental harm. Thus, the net loss to consumers from the 100th gallon, including the environmental harm, equals the gap between the MSC curve and the MB curve—$1.00. For the 99th gallon, the net loss is once again the gap between MSC and MB—now a little less than $1.00. Adding up the net loss for each gallon gives the area of the triangle; the area equals one half the base ($1.00) times the height (20), so the net loss to consumers is $100.

Charge Polluters a Price

For virtually all economists, including me, the solution is straightforward. The government must step in and assume ownership on behalf of the public, and then do what a typical private owner of a resource does: charge a price for its use. In other words, the market fails because a key element—ownership of a valuable resource—is absent. The solution is to restore the market by restoring the missing ingredient: ownership of the resource, and a price for its use.

There are two ways in which the government can charge a price: by

Figure 3.1 **Market Failure**

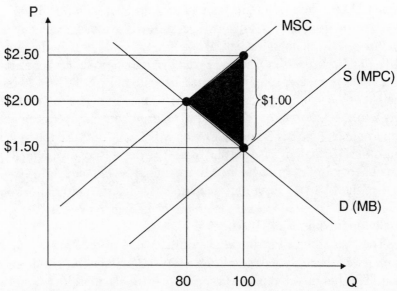

tax and by permit. Under the tax method, the government sets a tax per unit of pollutant X in a particular geographic region. Polluters are then free to respond; for each unit they pollute, they must pay the tax. Under the permit method, the government decides the aggregate quantity of pollutant X it is willing to tolerate in a particular geographic region. It then auctions that quantity of permits to polluters, where each permit allows the owner to emit one unit of pollutant X. Emission of a unit of the pollutant without a corresponding permit would be illegal. The permit price would be set by supply and demand.

Note the difference between the two methods. Under the tax method, the government fixes the price per unit of pollutant X, but the response of polluters determines the aggregate quantity of pollution. Under the permit method, the government fixes the aggregate quantity of pollutant X, but the bidding of polluters determines the price of a permit—hence, the price per unit of pollutant X.

Which method is better? It depends. We'll return to this question shortly. But note this: both methods raise government revenue so the government will need less revenue from other taxes to balance its budget.

Let's return to the gasoline market, and once again assume that the

Figure 3.2 **Charge Polluters a Price**

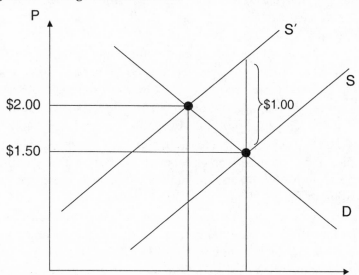

environmental harm per gallon is $1.00. Then economists recommend that sellers of gasoline be charged $1.00 per gallon—implemented either by levying a tax of $1.00 per gallon, or by requiring the purchase of a permit, at a price of $1.00, for each gallon sold. The tax or permit of $1.00 per gallon raises the private cost per gallon to sellers by $1.00, so it shifts up the supply curve in Figure 3.2 by $1.00. The result is a rise in the price of gasoline, a reduction in the quantity of gasoline actually used, and therefore a reduction in pollution. Now the "corrected" market generates the socially optimal amount of gasoline—80 gallons.

Charging polluters a price forces them to *internalize the externality.* Previously, the harm they were causing the environment was "external" to them because they did not have to pay for it. By charging a price, the government makes the cost "internal" to the polluters— they must now pay for using up a valuable resource—the environment—just as they must pay for using up other valuable resources like labor and materials. Like any other internal cost, they try to pass it on to buyers by raising the price of gasoline, which cuts down on the purchase of gasoline and pollution, as shown in Figure 3.2.

Note that, as shown in Figure 3.2, the price of gasoline rises less

than the tax. In this example, the price rises from $1.50 to $2.00, or $0.50, so consumers bear a burden of $0.50 per gallon. The gas station owners also bear a burden of $0.50 per gallon. Why? They charge drivers $2.00 per gallon, but then must send $1.00 per gallon to the government, so they keep only $1.00 per gallon, which is $0.50 less than if there had been no tax. In this example, the burden of the tax is divided half and half—$0.50 on the buyer, and $0.50 on the seller. The split will not always be half and half. How the burden is split depends on the slopes of the D and S curve. For example, if the D curve were steeper (less elastic), then the price would rise above $2.00, and the buyers would bear more than half the burden of the $1.00 tax.

But aren't the drivers, not the sellers of gasoline, the actual polluters? Technically, yes. But it is more practical for the government to charge the business firms that sell gasoline, collect the revenue from them, and let the firms pass on their new cost to drivers by raising the price of gasoline. Drivers end up paying a price for gasoline that is $1.00 more per gallon than the regular marginal cost (the height of the initial supply curve at the quantity actually sold). It is easier to get business firms to send checks to the government than it is to get drivers to send checks. It is also much easier to audit business firms than drivers to make sure $1.00 is being sent to the government for each gallon. So when we say charge "polluters" a price, we mean charge the business firms that either directly pollute, or enable pollution by selling a product that generates pollution when it is used.

Before leaving this example, let's consider an important issue. Contrary to our assumption that all kinds of gasoline generate the same pollution per gallon, different kinds of gasoline generate different amounts of pollution per gallon. Given this fact, should the government vary the tax per gallon according to the amount of pollution? The answer is yes. By varying the tax per gallon, producers of gasoline are given an incentive to produce the kind of gasoline that involves less pollution. In other words, there should be some tax per unit of pollution, not per gallon of gasoline.

Consider another example where it is very important to levy the tax per unit of pollution, not per unit of product: sulfur dioxide from the production of electricity. Whenever the generation of electricity involves burning coal containing sulfur, sulfur dioxide is produced.

If this sulfur dioxide is emitted into the air, "acid rain" is created with harmful effects. But it would be a mistake to levy a tax per unit of electricity. Why? Because the producers of electricity—electric utilities—can vary the amount of sulfur dioxide that gets emitted per unit of electricity produced. The utilities can use low-sulfur coal instead of high-sulfur coal, or even switch to a fuel that contains no sulfur. The utilities can install "scubbers" which capture most of the sulfur dioxide so that it is not emitted into the air. It is important to give electric utilities an incentive to make choices that reduce the amount of sulfur dioxide emitted into the air per unit of electricity produced. A tax per unit of electricity won't provide such an incentive, but a tax per unit of sulfur dioxide emitted will. Thus, the government should levy a tax per unit of sulfur dioxide emitted into the air, not a tax per unit of electricity produced, just as it should levy a tax per unit of gasoline pollution, not a tax per gallon of gasoline.

We can state our general conclusion: Charge polluters a price per unit of the pollution they emit, not a price per unit of the product they produce. In the remainder of this chapter, we will always assume that the price charged—either through a tax or a permit—is per unit of pollution, not per unit of product.

Facing the Trade-off

Now imagine a violently lethal pollutant. Even the slightest bit of it would cause enormous casualties. What do economists say about that? The same thing as any other sensible citizen: Ban it. But we economists insist on viewing a ban as an extreme case of our two pricing methods. Under the tax method, the more harmful the pollutant, the higher should be the tax per unit of pollutant; in the extreme, the tax should be so high that no polluter can afford to emit even a single unit. Under the permit method, the more harmful the pollutant, the smaller should be the aggregate number of permits that are auctioned; in the extreme, the number auctioned should be zero. So economists have no problem with an extreme case.

But we insist that society face up to the basic trade-off shown in Figure 3.3. Getting more environmental quality involves giving up some material output. For example, suppose that under a free market with no governmental environmental policy, the economy would go

Figure 3.3 **Output Versus the Environment**

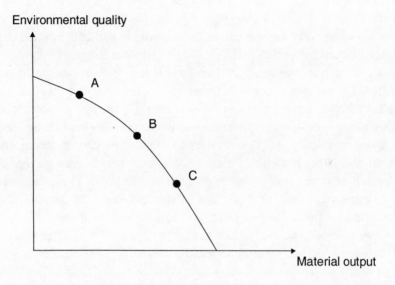

to point C. How can environmental quality be raised? One way is for firms to cut back on material output involving pollution. Another is for firms to switch to cleaner but more costly production techniques. Still another is for firms to produce and pollute exactly the same but have the government use resources to treat the pollution. However environmental quality is improved, there will be less material output for consumers to enjoy. Hence, the economy will move to point B indicating less material output. To raise environmental quality still higher, still more material output must be sacrificed, so the economy would move to point A.

All that economists ask is that people face up to the trade-off. Consider again a violently lethal pollutant. Here's what economists would advise. Imagine allowing a single unit to be emitted. Estimate the harm done. Then estimate the additional material output that the emission would make possible. Compare the two. If you decide that the harm outweighs the benefit from the material output, then by all means ban the pollutant.

But now consider a less harmful pollutant. Suppose that emitting a single unit would not do much damage, but would allow a highly valued increase in material output. If the public decides that the harm

from the extra pollution is less than the benefit from the extra output, then by all means allow this unit to be emitted. And now make the same comparison for a second unit. If the harm is still less than the benefit, then make the same comparison for a third unit.

As the amount of pollution rises, the harm from another unit of pollution is likely to rise, and the benefit from the associated output is likely to fall. So at some point, the harm from another unit will at last exceed the benefit. Clearly, pollution should be allowed up to this point, and no further. So at last we have arrived at the socially optimal amount of pollutant X. We've properly balanced the benefit from each unit of pollutant X against its harm or cost.

It's now safe to tell you something. You've just gone through an exercise in *marginal analysis*—the most fundamental technique in microeconomics. Marginal analysis means finding an optimum by reasoning one unit at a time. Did it seem simple, commonsensical? Don't be alarmed, but you may be an instinctive economist.

I hope you're thinking, "It sounds nice in theory, but isn't it hard to do in practice?" Absolutely. But it would be a great step forward if people realized that there is such a thing as a socially optimal amount of each pollutant, that except in an extreme case it is greater than zero, and that in theory we can locate it by marginal analysis—reasoning one unit at time.

Now to some practical problems. Air or water quality is a "public good." This means that it is impossible to improve the quality for me without also improving it for my neighbor. Yet I may care, and he may not. The same is true of national defense, another public good, but not true of a TV, which is a typical private good. I can easily get a higher-quality TV while my neighbor sticks with his little black-and-white model. But not so for environmental quality or national defense.

Imagine that the government conducts a survey. An interviewer asks each citizen: "Be honest, how much would you pay to raise air quality from grade C to grade B?" Then the government sums the amount all honest citizens would be willing to pay, and compares it to the cost—the lost material output. If the sum of payments exceeds the cost of lost output, air quality should be increased to grade B. And then the government asks the same question about raising it from grade B to grade A. Once the sum of what people are willing to

pay no longer exceeds the cost of another unit of quality, the government has arrived at the socially optimal grade of quality.

But there's a little practical problem. How can the government get citizens to be honest? After all, what would you think if a government interviewer came to your door and asked, "How much are you willing to pay for Z?" You might think, "If I tell him I'd be willing to pay $100 for Z, that's actually what he'll force me to pay." I'm sure *you* would be honest, but I'll bet you know someone who would understate his true willingness to pay.

Economists have tried to invent ingenious techniques to induce citizens to answer honestly. But although some progress has been made, such techniques are not often used. So we must admit that determining the socially optimal amount of pollutant X will be as imperfect, in practice, as determining the optimal amount of national defense or police protection. Imperfectly, the legislature must simply do the best it can to weigh cost against benefit.

Why Economists Advocate Pollution Prices

What's so great about using prices for pollution? Once the target for pollutant X in a particular region is set, why not just assign a quota (ceiling) to each polluter of X, to ensure that the aggregate quantity of X equals our target?

Now we come to the heart of the matter. Once the aggregate target has been set, how do we decide which firms should do the polluting—that is, how do we "allocate" pollution among the polluters? For example, suppose our aggregate target for pollutant X in region R is 1,000 units, and there are 100 polluters of X in the region. One simple approach would be to allow each polluter to emit 10 units of X. Another approach would be for the government to mandate specific low-pollution production techniques for the polluters of X. In fact, this "technology-forcing" approach is generally taken by the U.S. government. Unfortunately, each of these approaches would be a poor way to handle the allocation problem. Why?

First of all, the 100 polluters of X produce a variety of products. Consumers value some of these products more than others. Surely we want the allocation to take account of consumer preference for the products. Second, the polluters differ in technological options.

Some can cut back pollution easily with little additional cost. Others must incur a substantial cost increase to achieve the same reduction. Since polluters will pass on cost increases to consumers, we want the allocation to take account of these technological options.

It might seem that government agents could interview consumers about their preferences for products, and the polluters about their technological options. But these interviews would be costly and of dubious value. It's far from obvious what questions to ask consumers. And while the questions are clearer for the polluters, why should they tell the truth? Wouldn't any polluter try to exaggerate the cost of cutting back in order to win a higher ceiling?

I'm sure you've realized by now that interviews could also be used to decide how much to produce of any product, pollution aside. Fortunately, we don't use them. Instead, our price system gets the job done. Both consumers and producers respond to prices, and there is no need for government interviews.

Let's see how pollution prices would handle the allocation problem. First, consider the tax method. Suppose the government sets a tax of $20 per unit of pollutant X. The polluter whose products have good substitutes will reason: "I can't afford to pay this tax on many units, because when I try to pass the cost on to my customers by raising my price, they will simply shift to substitutes." By contrast, the polluter whose products are highly valued by consumers will reason: "I can afford to pay the tax, because when I try to pass the cost on to my customers, I will succeed." So who ends up cutting pollution sharply? The polluter whose product has good substitutes. And who ends up cutting back relatively little? The polluter whose product is highly valued by consumers. And this is exactly the pattern of cutback we want.

Next, consider the polluter with technological options that enable a reduction in pollution at little additional cost. He reasons: "Rather than pay the tax, it is cheaper for me to switch technologies and reduce pollution." By contrast, consider the polluter with few technological options. Only at high cost can he reduce pollution. He reasons: "I'm still better off paying the tax, because it would be even more expensive for me to switch technologies." So who cuts back pollution sharply? The polluter with good technological options. And who cuts back pollution relatively little? The polluter with few techno-

logical options. And this is exactly the pattern of reduction we want.

Of course, when each polluter of X decides how much to pollute if the tax is $20, it may turn out that aggregate pollution of X in the region will be 1,200 or 800 instead of the target, 1,000. If pollution is 1,200, then the government should raise the tax above $20; if pollution is 800, then the government should lower the tax below $20. Eventually, the government will find the tax that approximately achieves the aggregate target of 1,000.

Now consider the permit method. The government would auction off 1,000 permits, so the supply of permits would be fixed at 1,000. The government would get polluters to reveal their demand for permits by asking them to place orders for permits at each of the following prices: $10, $20, and $30. The polluters would submit their orders at each price. For example, at a price of $10, polluters might order (demand) 1,200, while at a price of $30, polluters might order 800. Suppose that at a price of $20, polluters would want to order 1,000. The government would then announce that the permit price is $20 and would sell each polluter the number of permits it ordered at that price.

If the price of a permit is $20, then a $20 per unit pollution tax should produce exactly the same pattern of pollution reduction across polluters and yield aggregate pollution equal to 1,000. After all, a polluter doesn't care whether the $20 price per unit is called a permit price or a tax. He will figure his profit and do the same thing. Therefore, whether the price is charged through the tax or permit method, the desirable pattern of cutback across polluters is induced. So we see how a price system results in the socially optimal allocation of pollutant X across polluters. Is some more general principle at work here? Of course. A price system results in the socially optimal allocation of any resource across users—whether the resource is labor, capital, materials, or pollutant. One requirement of being an economist since Adam Smith is to understand this point. And that is why economists want to use prices to allocate pollution among polluters.

There is another way to put this. We can fight over what the target for pollutant X should be. Should it be 800 units or 1,200 units instead of 1,000? I've already admitted that it is tough to answer this question in practice, and economists don't claim to offer any good way of selecting the target. But once the target has been set, we can

surely agree on this: Let's achieve the pollution target with the minimum sacrifice in material output.

That is exactly what pollution prices can do. Pollution prices induce a socially desirable pattern of pollution reduction across polluters. This means that the target is achieved with the minimum loss in the value of material output.

The Passionate Objector

But alas, a passionate objector has risen to his feet and will keep silent no longer. He cries, "Isn't a pollution price a 'license to pollute' and therefore reprehensible?" I'm afraid, my passionate friend, that you've confused two distinct tasks: First, what should be the aggregate target? Second, given the target, how should we allocate the pollution among polluters? Prices apply only to the second task, while you are really concerned about the first task—the setting of the target. No doubt you want a target of zero. Fine; make your case, and we will listen. If you succeed, then we will simply ban the pollutant. But if you fail to convince us and we decide to tolerate a certain amount of pollutant X, then surely even you will agree that we should achieve the target with the minimum sacrifice in material output. And that's where prices come in.

Our passionate objector shifts his ground and asks, "Won't the polluters just pass on these pollution prices to me, an innocent consumer, by raising product prices to cover these charges?" Yes, answers any honest economist, they certainly will. "And," continues our righteous objector, "why should I pay for their foul pollution? Let them pay for it!"

I'm afraid, my passionate objector, that once again you have missed the point. The whole object of pollution prices is to confront consumers with the environmental cost of the products they buy. The price system is an information system. The price of each product is supposed to convey information to the consumer—namely, the cost of producing it. If the price is less than cost, then the consumer is misled and demands too much of the good; if the price is greater than cost, then the consumer demands too little of the good. The free market fails because the price of high-polluting (HP) output is too low relative to the price of low-polluting (LP) output. Why is the price of

HP output too low? Because it does not include the environmental cost. And why not? For the simple reason that polluters are not charged for polluting. And why aren't they charged? Because there is no owner of air or water to impose the charge. So consumers are induced, by the false price signals, to consume too much HP output and too little LP output. The whole point of pollution prices is to raise the price of HP output relative to the price of LP output, so that consumers receive accurate information, and as a result shift consumption from HP to LP products.

"But is this fair?" replies our objector. "Is it fair that I, an innocent consumer, should pay, instead of the dirty polluters?"

I beg your pardon, but what do you mean, innocent? You are enjoying a product that entails pollution. Yes, the producer did the dirty work, and now you want to enjoy the product without any responsibility for what its production required. How admirable!

Also, to whom do you want to shift the burden? The inanimate corporation? Alas, there is a little obstacle to your strategy, and it is this: Business firms don't bear burdens; only people do. The people may be consumers, workers, stockholders, or managers. But it is ostrichlike to hope that lifeless corporations, not flesh-and-blood people, will bear the burden. So the debate is really about which people should bear it. Most economists take this position: Let each consumer pay a price that reflects the cost of the product, including the environmental cost, so that the price system conveys accurate information.

Somewhat subdued, our objector at last asks a better question. "But aren't there some practical problems with your two pricing methods that you've been concealing?"

Well, *concealing* isn't the right word. But I'll admit I haven't gotten around to them. So let's consider a few problems. Perhaps the most important practical problem is this: Pricing requires metering—measuring each polluter's emissions—while mandating low-pollution technologies does not. In some cases, metering may be too costly or unfeasible. Naturally, economists recommend using pricing only if the cost of metering is less than the benefit of pricing.

The tax method has another problem: It can never guarantee that the pollution target will be met precisely. In response to a given tax per unit, polluters may emit too much or too little. If too much, the

government can raise the tax; if too little, the government can lower it. But the government may never hit it exactly right. Of course, we should keep this problem in perspective. Earlier, we saw that selecting the target is extremely difficult in the first place. No selected target should be regarded as sacrosanct, because it surely differs from the social optimum. So moderately missing the target should not be viewed with alarm, because the target itself is probably not the social optimum.

While the permit method seems to guarantee that the target will be achieved, it has other problems. Will the auction occur on a single day for the year? What if a firm decides it needs more or fewer permits as the year progresses? Will the permits be transferable? Will there be a continuous resale market for permits? Who will be allowed to bid for permits? Will firms buy permits simply because they expect to sell the permits at a higher price, or because they want to keep competitors from getting the permits they need to produce? The tax method has none of these problems.

Thus far, neither the United States nor any other nation has relied heavily on pollution prices to implement environmental policy. Should economists surrender to despair? Not at all. The cause of pollution prices has slowly begun to make progress. Some recent experiments with pollution prices (in the form of transferable permits) give grounds for hope. Nevertheless, we must confess that governments still generally mandate specific production techniques for polluters. So we continue to achieve a given level of environmental quality with an unnecessarily large sacrifice in material output.

Faster Economic Growth and Pollution

Will faster economic growth doom us to more pollution? Not necessarily. Suppose that today, 100 units of output are accompanied by 10 units of pollution. Isn't it inevitable that 200 units of output will be accompanied by 20 units of pollution? Not necessarily. The key to breaking the link between output and pollution is substitution of less-polluting goods for more-polluting goods, and of low-pollution production technology for high-pollution production technology.

Material output consists of many goods and services. With today's technology, some are high-pollution (HP) and others are low-pollution

(LP). Even if technology remained the same, we could raise our total output without raising pollution if we shifted the composition of our output from HP to LP goods.

How can we induce such a shift? I'm sure you can guess: by gradually raising our set of pollution prices as our output grows. This will raise the price of HP goods relative to the price of LP goods, and consumers will be induced to shift their demand from HP to LP goods. Producers will therefore be compelled to shift production from HP to LP goods.

Moreover, production technology can be shifted as output grows. Many goods can be produced using alternative methods that generate different amounts of pollution. We could raise our total output, without increasing pollution, by shifting from high- to low-pollution production technologies, using pollution prices as a deterrent against the use of HP technology. Such prices would make it profitable for firms to reduce the pollution that accompanies output by shifting to low-pollution production technologies.

So, can we grow faster without reducing environmental quality? Yes. But to do so, we must gradually raise our set of pollution prices as our output grows, thereby inducing these substitutions as growth takes place.

It is true that as output grows, it gets "harder" to maintain a given level of environmental quality in this sense: If nothing were done to change the mix of HP and LP goods, or the mix of HP and LP production technologies, then environmental quality would deteriorate. All the more reason for using pollution prices (through taxes or auctioned permits), rather than mandated production techniques, to implement environmental policy. The more importance we attach to raising output per person, the more important it is to maintain environmental quality with the minimum sacrifice in material output.

Faster Growth and Depletion

Will faster growth be stalled by resource depletion? Again, not necessarily. Suppose that today, 100 units of output are accompanied by the depletion of 10 units of a natural resource. Isn't it inevitable that 200 units of output will be accompanied by the depletion of 20 units of the natural resource? Fortunately, not necessarily.

Would it surprise you to learn that the key to breaking the link between output and natural resources is substitution of goods whose production requires low amounts of natural resources (LR goods) for goods whose production requires high amounts of natural resources (HR goods), and of LR production technologies for HR production technologies?

Material output consists of many goods and services. With today's technology, some are *high resource* (HR) using, and others are *low resource* (LR) using. Even if technology stayed the same, we could raise our total output, without raising depletion, if we shifted the composition of our output from HR to LR goods.

How can such a shift be induced? I'm sure you can guess: by a rise in natural resource prices. If this happens, the price of HR goods will rise relative to the price of LR goods, consumers will be induced to shift demand from HR to LR goods, and producers will therefore be compelled to shift production from HR to LR goods.

Moreover, production technology can be shifted as output grows. Many goods can be produced with alternative technologies that utilize different amounts of natural resources. We could raise our total output, without raising depletion, by shifting from HR- to LR-using production technologies.

How can such a shift be induced? I'm absolutely sure you can guess: by a rise in natural resource prices. If this happens, it will be profitable for firms to reduce the depletion that accompanies output by shifting to LR-using production technologies.

But now we come to an important difference between pollution and depletion. The government sets the pollution tax or the volume of permits for auction, so it is up to the government to raise the tax or limit the volume of permits (resulting in a higher permit price) as output grows. The government must act because of the absence of private ownership of air and water. But there is private ownership of natural resources and it is the market that sets resource prices. So the key question becomes: Will the market raise resource prices as output grows?

The best way to know the answer is to imagine, pleasantly, that you are a resource owner. Your resource lies underground, and you have the option of mining it and selling it to firms for production today or holding it in the ground to await a future price. What do

you do? How much do you mine and sell, and how much do you keep underground?

Suppose you read that scientists are on the verge of a breakthrough that will provide a cheap, abundant substitute for your resource. As a citizen, you are overjoyed at the advance of mankind through brainpower. As a resource owner, you are desperate. There is no point holding your resource underground to await a future price. So you mine and sell at full capacity.

Now suppose that instead, you read that scientists regard such a breakthrough as unlikely, so your resource, growing scarcer each year, will command a rising price in future years. As a citizen, you are in despair over the failure of brainpower to advance mankind. As a resource owner, you can hardly conceal your glee. Why sell to the market today when a much higher price awaits you in the future? Even if you need cash, it would be better to borrow than to foolishly mine and sell at full capacity. So you hold most of your resource underground, as do other owners of that resource. Its price rises immediately, inducing firms to cut down on current utilization.

But look what has happened. Lo and behold, conservation has been achieved. You and your fellow resource owners have cut down mining and selling, and kept most of your resource underground. Of course, your motive was not exactly noble. In fact, the word "conservation," with its public-spirited connotation, never entered your mind. In truth, the word "profit," not conservation, was on your smiling lips as you fell asleep each night.

But old Adam Smith would not have been surprised. In your pursuit of profit, you advanced the public welfare. With the threat of your resource becoming scarce in the future, the market gave you a profit signal to conserve now. And so you did. Immediately, your cutback in mining and selling raised your resource's price and induced a cutback in its utilization. We did not have to rely on your benevolence, but only your self-interest, to slow the depletion of your resource. So without any government action, the market will automatically tend to generate the desired rise in the price of a resource as output grows if the resource is really becoming scarce.

We haven't even mentioned another important consequence of the rise in current prices and anticipated future prices: You and your fellow miners will go exploring. With high prices, more exploration is

worth it. Also, scientists will go to work trying to invent substitutes. With high prices, there's a lot of money to be saved in finding alternatives. So not only do high prices cut current use, they also create incentives for long-term solutions.

Can it really be this simple? Unfortunately, no. The market can make mistakes. After all, no one has a crystal ball. Everyone may underestimate the difficulty of finding substitutes for your resource. So everyone may underestimate the future price of your resource. You and other owners may decide to mine and sell too much today. In the future, everyone may regret it.

So it would be naive to believe that the market handles the depletion problem perfectly. But it should be some consolation that the market has the tendency to conserve any resource that threatens to grow scarce in the future, and to induce the invention of substitutes.

4

International Trade

This is a tale of two islands: Labor Land and Capital Land. The islands were separated by a sea, and before the invention of ships, neither knew of the other's existence. In fact, each thought it was the only land on earth.

Labor Land never accumulated much machinery, and most Labor Landers acquired little education. Yes, most learned how to make some simple tools, grow a few crops, build simple houses, and read, write, and do arithmetic. But that was all. Naturally, their productivity, wage, and standard of living were low. However, a few Labor Landers owned machines and were highly educated and trained; naturally, their productivity, wage, and standard of living were high.

Things were very different in Capital Land. Capital Land had lots of machinery. Capital Landers invented productive machines and discovered how to make many products. Most were highly educated and trained. Naturally, their productivity, wage, and standard of living were high. However, some Capital Landers failed to own capital and obtain much education and training. A few were simply lazy, but others, despite their best efforts, were unable to learn advanced skills. Naturally, their productivity, wage, and standard of living were low.

One day there was great excitement at a huge factory located on the Capital Land seashore. A huge object, constructed in the factory, was being towed into the water. A crowd of people gathered to watch.

"It's going to sink, it's going to sink!" shouted a little boy. But it didn't sink. To the disbelief of children and adults alike, it floated. A cheer arose from the workers in the factory, then another from the crowd. The factory manager climbed up a ladder on the outside of the object, then stepped over a railing, stood on a deck inside the object, and addressed the crowd.

"We've done it. I am proud to announce that all of you are witnessing a historic moment: the launching of the first ship. For thousands of years we have stared at the sea that surrounds our island. Now at last we will be able to discover if we are truly alone on earth."

One week later a brave crew set sail at dawn. As a crowd watched, the ship sailed slowly into the distance. As night fell the ship could no longer be seen. The next morning it was gone.

"I can't see it anymore," said a little girl. "Did it sink?"

"We don't know," answered her mother. "They have enough food to sail for a month. They've been given orders to turn back in two weeks. Let us pray for them."

Exactly one month later a cheering crowd greeted the returning ship. The sailors hurried down the ladders, bursting with excitement.

"We discovered another populated island!" they shouted. "The people were friendly as can be. We're going to trade, we're going to trade!"

"What does 'trade' mean?" asked a little girl.

"It means," a sailor replied, "that we'll exchange some products we've made for some products they've made."

Comparative Advantage

"Why should we trade?" asked Autarkas. "Do they make anything that we can't make?" The sailors were suddenly quiet. Finally one sailor admitted, "I didn't see any product we can't make. You see, most Labor Landers have little machinery, education, and training. Most of our workers, equipped with advanced machines, education, and training, can out-produce them in every product. I'll give you two examples. First, consider a product we are most proud of: hitech [pronounced 'high teck']."

"Can Labor Landers even make a hitech?" asked Autarkas.

"This may surprise you," replied the sailor. "They actually man-

age to produce a small quantity of hitechs. But it's pitiful. One of our workers can produce 10 times as many hitechs in the same time as one of their workers can."

"What's your second example?" asked Autarkas.

"It's the product lotech [pronounced 'low teck']," replied the sailor. "Lotech doesn't require a lot of machinery, education, and training to produce, so the Labor Landers do a lot better making lotechs than they do making hitechs."

"But," persisted Autarkas, "can one of their workers beat one of our workers at making a lotech?"

"Well," conceded the sailor, "not really. One of our workers can produce 5 times as many lotechs in the same time as one of theirs can."

"So," said Autarkas with a satisfied smile, "we have an *absolute* advantage in every product. Our advantage is 10 times for hitechs and 5 times for lotechs. So again, I repeat, why trade? We're better off making every product ourselves."

Just then there was a disturbance in the crowd as a young, well-dressed gentleman made his way to the front until he stood facing Autarkas and the sailor.

"Autarkas, what you say sounds so sensible," the young gentleman said. "At first glance it surely seems pointless to trade with Labor Land when we have an absolute advantage in every product. But, alas, what seems sensible at first glance does not always turn out to be true on deeper analysis. That analysis shows that Capital Land and Labor Land can both gain from specializing in production and then trading freely."

"Who are you?" asked Autarkas.

"I'm Ricardo," the young gentleman replied.

"I've heard of you," said Autarkas. "You made a fortune as a stock-broker as a very young man, and now you serve in our legislature. You are certainly a successful, practical man. So how can you think trading with Labor Land makes sense for us?"

"You were kind enough to note my practical successes. But you may not be aware that two years ago I published a treatise on political economy, the result of many hours of analysis and work. In one chapter of my treatise I explained why free trade would make both lands better off."

Table 4.1

Output Per Worker Per Day

	Hitechs	Lotechs
Capital land	10	20
Labor land	1	4

"But this is quite remarkable," said Autarkas. "How could you even write about trade with another land in your treatise? We just discovered the existence of Labor Land, and your treatise was published two years ago!"

"I have two answers," replied Ricardo. "First, the idea that specializing and trading is mutually beneficial applies to a single land, two lands, or many lands. Second, I admit that in my treatise chapter I did assume that there were two lands. But you see, I can assume what I want. I'm an economist."

"If you are really an economist, you are quite right: You can assume anything. But why will we gain from free trade with Labor Land?" repeated Autarkas.

"Sailor," asked Ricardo, "how many lotechs does one Labor Lander produce in a day?"

"He produces 4," replied the sailor, "while one of us produces 20 in a day. As I said before, we're 5 times as productive as they are in lotechs."

"And how many hitechs does one Labor Lander produce in a day?" asked Ricardo.

"Just 1," answered the sailor, "while one of us produces 10 in a day. As I said before, we're 10 times as productive as they are in hitechs."

"Let me display the numbers in a table," exclaimed Ricardo. And he did (Table 4.1).

"We're better at making both goods, so let's produce both hitechs and lotechs ourselves," said Autarkas.

"That would be a mistake," said Ricardo. "When one of our workers spends a day making lotechs instead of hitechs, he produces 20 lotechs instead of 10 hitechs. So our sacrifice of 10 hitechs gets us 20 lotechs to consume. Do you agree that it would be better if our worker

spent the day making 10 hitechs and then traded the 10 hitechs for 30 lotechs from Labor Land?"

"Obviously," replied Autarkas. "Of course we'd rather have 30 lotechs at the end of the day than 20 lotechs. But will Labor Land agree to the trade?"

"They should," replied Ricardo. "When one of their workers spends ten days making hitechs instead of lotechs, he produces 10 hitechs instead of 40 lotechs. So he must sacrifice 40 lotechs to consume 10 hitechs. Wouldn't he be better off producing and trading only 30 lotechs to get 10 hitechs?"

"Indeed he would," answered Autarkas.

"Note this," said Ricardo. "Within Capital Land, the choice is 1 hitech or 2 lotechs, while within Labor Land, the choice is 1 hitech or 4 lotechs. It follows that if Capital Land can trade 1 hitech for more than 2 lotechs, it would gain; and if Labor Land can obtain 1 hitech for less than 4 lotechs, it would gain. So trade at any ratio between 1 to 2 and 1 to 4—for example, 1 to 3—would cause both lands to gain. With trade, Capital Land gets 3 lotechs for every hitech it sacrifices; without trade, it gets only 2. With trade, Labor Land gets 1 hitech for every 3 lotechs it sacrifices; without trade, it must sacrifice 4. That's why the trade of 10 hitechs from Capital Land for 30 lotechs from Labor Land (1 hitech for 3 lotechs) benefits both lands."

"Can you draw me a diagram to show me how both countries would be better off?" asked Autarkas.

"Certainly," said Ricardo. And he drew the Capital Land diagram in Figure 4.1. "I happen to know that there are 2,000 workers in Capital Land, so from Table 4.1, if all workers made lotechs they could produce 40,000 lotechs per day, and if all workers made hitechs they could produce 20,000 hitechs per day. Of course, if half the workers made lotechs, and half, hitechs, they could produce 20,000 lotechs and 10,000 hitechs per day. In fact, it would be possible for them to produce any combination of lotechs and hitechs that lies on the line connecting point (40,0) and point (0,20). To save space, I've written 40 for 40,000, and 20 for 20,000."

"Why don't you call that line a *production possibilities line*," suggested Autarkas.

"That's a splendid suggestion," replied Ricardo. "But possible

Figure 4.1   **Specialization and Trade**

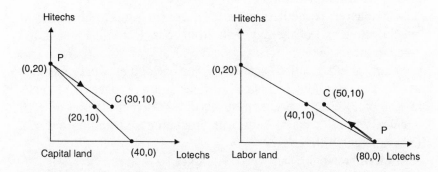

production points for a country don't always lie on a straight line, so I will call it a *production possibilities curve*, which in this case is a line."

"May I try to draw the Labor Land diagram?" asked Autarkas.

"By all means," said Ricardo. "It's always better for a student to try to draw a diagram himself—that really tests whether the student knows what's happening. But I must give you one fact before you can draw it. I happen to know that there are 20,000 workers in Labor Land."

Autarkas thought for a moment, and then drew the Labor Land diagram in Figure 4.1. Using Table 4.1, Autarkas figured out that they could produce point (80,0), or point (0,20), or any point along the line joining these two points.

"Excellent," said Ricardo. "Without trade, people in each country want some of both goods. It just so happens that without trade each country produces and consumes at the half-way point on its production possibilities line: Capital Land produces and consumes (20,10), and Labor Land produces and consumes (40,10); this doesn't have to be true, but it just happens to be the case in these two countries."

"How nice for our diagrams," said Autarkas.

"By the way," continued Ricardo, "don't forget that there are ten times as many workers and people in Labor Land (20,000 vs. 2,000), so consumption per person is much lower in Labor Land even though Labor Land produces more lotechs (40 vs. 20) and the same number of hitechs (10 vs. 10)."

"I'm glad you reminded me," said Autarkas.

"Now, Autarkas," said Ricardo, "let's see if you can show what happens if Capital Land specializes in producing hitechs, Labor Land specializes in producing lotechs, and they trade 10,000 hitechs for 30,000 lotechs."

Autarkas thought for a moment, and then said, "Capital Land would produce at point (0,20). It would then export 10,000 hitechs and import 30,000 lotechs so it would consume at point (30,10)."

"Exactly right," exclaimed Ricardo.

"Labor Land," continued Autarkas, "would produce at point (80,0). It would then export 30,000 lotechs and import 10,000 hitechs so it would consume at point (50,10)."

"So," exclaimed Ricardo, "with trade each country goes to its respective corner like two boxers before a match, and produces at its point P."

"I see," said Autarkas. "Capital Land produces no lotechs and 20,000 hitechs, while Labor Land produces 80,000 lotechs and no hitechs."

"Then," continued Ricardo, "each country trades, advancing from its corner P toward the center of the ring to its point C which indicates the amount it consumes of each good after trade."

"I notice," said Autarkas, "that the slopes of the two trade lines, PC, are the same for both countries: 1 hitech for 3 lotechs. Do the slopes of the two trade lines have to be the same?"

"Of course," smiled Ricardo. "The two countries are trading with each other, so if Capital Land exports 1 hitech and imports 3 lotechs, then Labor Land is obviously importing 1 hitech and exporting 3 lotechs."

"Now," asked Ricardo, "are both countries better off with trade than without it?"

"Why, yes, they are," exclaimed Autarkas. "With trade, Capital Land consumes (30,10), which is clearly better than the (20,10) it consumed without trade. And with trade, Labor Land consumes (50,10), which is clearly better than the (40,10) it consumed without trade. That's simply remarkable," said an astonished Autarkas.

"It is indeed," said Ricardo. "I can't tell you how excited I was when I discovered it. If we in Capital Land are 10 times as productive as Labor Land in hitechs but only 5 times as productive as Labor Land in lotechs, then of course we have an absolute advantage in

both goods. On average, we're 7.5 times as productive. But I propose to say that we have a *comparative* advantage in hitechs because our productivity multiple (10) is above average (7.5) in hitechs, and Labor Land has a *comparative* advantage in lotechs because our productivity multiple (5) is below average (7.5) in lotechs. We can both benefit if we each specialize in producing the good in which we have a comparative advantage, and then trading it."

"But the way you've defined comparative advantage," said Autarkas, "each land must have a comparative advantage in one of the two goods."

"Correct," replied Ricardo.

"So you are claiming that mutually beneficial trade is always possible, even when one land has an absolute advantage in all goods," said Autarkas.

"Exactly," replied Ricardo.

Free Trade

"But," said Autarkas, "you have not shown that free trade will actually benefit both lands. You've only shown that a mutually beneficial trade could be negotiated by planners, not that it would come about automatically in a free market."

"You are quite right," replied Ricardo. "So let me now demonstrate that in a free market, Capital Land will specialize in producing hitechs and export them and Labor Land will specialize in producing lotechs and export them, and these actions under a free market will result in both lands being better off."

"You've got quite a task," said Autarkas. "Labor Land uses money called labs, while we use money called caps. Their prices are in labs while ours are in caps. Before we can trade, we have to know what the exchange rate is between labs and caps. They want to sell me goods priced in labs and they want to be paid in labs. I need to know how many labs I can get at a bank for each cap—that is, I need to know the exchange rate—before I can figure out whether I need less caps to buy a lotech made in Labor Land than I would need to buy a lotech made in Capital Land. I need to know which source is cheaper. You're going to have to construct a very complicated numerical example, and while I might understand it, the crowd will never follow it."

"Not at all," replied Ricardo cheerfully. "I can do it without a numerical example. Suppose that at the initial prices and exchange rate, consumers in both lands would want to buy all their goods from Capital Land. In other words, when each consumer figures out which source is cheaper, it turns out that both goods made in Capital Land are cheaper."

"Sounds great," said Exportas, who had been listening to the whole discussion. "We'll be able to export hitechs and lotechs and make lots of money."

"You're a fool," interjected Importas, who was standing right next to him. "We'll be sending lots of goods to Labor Land for their people to consume, while all we'll get are pieces of paper called money."

"Actually," said Ricardo, "there's no point in you two getting into one of your famous arguments, because the situation won't last. If everyone tries to buy all goods from our producers, our prices and wages will rise, and if no one tries to buy goods from their producers, their prices and wages will fall."

"What will happen next?" asked the sailor.

"That's easy," replied Ricardo. "At some point, our rising prices and their falling prices will make one of our two goods more expensive. At that point, consumers will decide to buy that good from Labor Land, while still buying the other good from us because it's still cheaper. So the Labor Landers will specialize in making the first good, and we will specialize in making the second good."

"Which of our two goods will become more expensive?" asked the sailor.

"See if you can guess," said Ricardo. "Remember, we're 10 times as productive making hitechs and only 5 times as productive making lotechs."

"Then it must be our lotechs that will become more expensive than their lotechs, even while our hitechs are still cheaper than their hitechs," said the sailor.

"Exactly," Ricardo smiled.

"So," said the sailor, "we'll specialize in producing hitechs, consuming some and exporting the rest, while they'll specialize in producing lotechs, consuming some and exporting the rest."

"Now notice something," said Ricardo. "We could have predicted the free trade outcome by checking comparative advantage. We're

10 times as productive in hitechs but only 5 times as productive in lotechs, so we have a comparative advantage in hitechs and they have a comparative advantage in lotechs. Sure enough, if we open up free trade, prices will adjust so that we each specialize in producing and exporting the good in which we have a comparative advantage. And that makes both lands better off."

"But wait," said Autarkas. "The trade must be at a ratio between 1 to 2 and 1 to 4—for example, 1 hitech for 3 lotechs (10 hitechs for 30 lotechs)—to make both better off. You haven't shown that."

"You are difficult to satisfy, Autarkas. I like that. You'd make a good economist. Alas, that would indeed take a numerical example. But the example would show that when prices adjust under free trade, the ratio would in fact end up somewhere between 1 to 2 and 1 to 4, so both lands benefit."

"I have another objection," said Autarkas. "You assume prices will rise in Capital Land and fall in Labor Land until one good from each land becomes cheaper. But what if prices are slow to adjust? What then?"

"That's easy," replied Ricardo. "Suppose prices are absolutely fixed. Initially, everyone tries to buy our goods and no one tries to buy goods made in Labor Land. So at the banks, Labor Landers try to exchange labs for caps so they can buy our goods—they supply labs and demand caps—but no Capital Lander tries to exchange caps for labs. So at the banks, there is demand for caps but no supply of caps. Suppose the exchange rate was initially 1 lab for 1 cap. With caps in short supply but high demand, the banks will change the exchange rate. They may now require 2 labs for 1 cap. The cap *appreciates* and the lab *depreciates* in value—now 2 labs, not 1, are needed to get 1 cap."

"But now," said the sailor, "our goods are more expensive to Labor Landers. Even though the price in caps is fixed, the price in labs has doubled."

"Very good," said Ricardo.

"And Labor Land goods are cheaper to us," continued the sailor, "because even though the price in labs is fixed, the price in caps has halved."

"Exactly," said Ricardo. "So even if prices are fixed, a flexible exchange rate will make our goods more expensive to Labor Landers,

and their goods cheaper to us. Eventually, one of the two goods will be more expensive when we make it and cheaper when they make it. So everyone will try to buy that good from Labor Land, not from us. Of course, that good will be a lotech. Now, at the banks, we will be supplying caps and demanding labs so we can buy their lotechs, and Labor Landers will still be supplying labs and demanding caps to buy our hitechs. At just the right exchange rate, the demand for caps will equal the supply of caps (and the demand for labs will equal the supply of labs) and the banks will have no reason to change the exchange rate."

"So," said the sailor, "once again they will specialize in lotechs and we will specialize in hitechs."

"That's right," said Ricardo. "So it doesn't matter whether prices rise and fall, or the exchange rate changes, or there is a combination of both—the result will be the same. Free trade will result in each land specializing in the good in which it has comparative advantage."

Fear of Trade

The sailors did not need to hear Ricardo to favor free trade. They were all for it because they would clearly profit from it. But people in Labor Land and Capital Land were another matter. Fear swept over Labor Land as soon as the ship from Capital Land left to return home.

"I've heard they're more productive in everything," said a frightened Labor Lander. "It is said they are 10 times as productive in making hitechs, and 5 times as productive in making lotechs. If we try to trade with them, they'll overpower us! We'll never be able to compete in anything, to sell anything. Free trade with Capital Land will be a disaster!"

Fortunately, when the second ship from Capital Land arrived in Labor Land, Ricardo was aboard. Immediately, he went directly to a TV studio where Labor Land's most popular talk show was in progress. As soon as Ricardo was introduced, the talk show host demanded that he explain why Labor Land wouldn't get slaughtered by trading with a productive powerhouse like Capital Land. The talk show audience was in a frenzy, and for a moment Ricardo was frightened.

"Don't worry," whispered the host. "The people we let in are always that way. It's good for our ratings. People love to watch hysterical people on TV."

"How interesting," whispered Ricardo to the host. "It's the same in Capital Land. The only difference is that the quality of our TV picture is much better than yours, so hysterical facial expressions can be seen more clearly by viewers. I suppose that's something you can look forward to under free trade."

Ignoring the shouting and frenzy, Ricardo calmly repeated the explanation he had given in Capital Land. "Suppose for a moment you can't sell anything because your prices are too high, and you want to buy everything from Capital Land because its prices are lower."

"Yeah!" screamed the audience. "Our factories will all shut down and we'll all be out of work!"

"Not at all," said Ricardo. "With everyone trying to buy goods made in Capital Land, its prices will rise. With no one trying to buy goods made in Labor Land, your prices will fall. This will all happen very quickly. Soon, one of your goods will be cheaper, and people in Capital Land will want to buy it. That good will be lotech, and you will specialize in its production, sell it to Capital Landers, and use your earnings to import hitechs. You'll do very well and be better off than you were before trade began. Now would you like me to explain how I know you will succeed at lotechs rather than hitechs?"

"No," said the talk show host, "that won't be necessary." No one in the audience was screaming anymore. In fact, half were yawning, and half had fallen asleep.

"But I've got a smashing numerical example," Ricardo said with enthusiasm.

The talk show host saw a frantic cut sign from the control room. He led Ricardo by the hand off the stage. Later, he learned that his show's ratings had plunged drastically during the five minutes that Ricardo had been speaking. True, many TVs remained tuned to his station, but only because the viewers had fallen asleep.

Nevertheless, Ricardo offered to give more lectures and numerical examples, but the editorial in Labor Land's leading newspaper spoke for the nation: "We'll trade, we'll trade, but please go home, Mr. Ricardo!"

And so Ricardo returned home. Upon arriving in Capital Land,

he found that a remarkable change had occurred. Initially, trade with Labor Land had been considered pointless. As Autarkas had said, "We're more productive in everything. Why trade?" Ricardo had persuaded Autarkas that Capital Land would benefit from trade with Labor Land, despite its absolute advantage in all goods. But now, to Ricardo's surprise, he found that fear of trade had spread among Capital Landers. He immediately went on a TV talk show to find out why.

"Our audience is very angry with you, Mr. Ricardo," whispered the host. "That's why we're having you on. Audience anger is great for our ratings."

"What seems to be the trouble?" Ricardo asked the audience.

"I'll tell you what the trouble is!" shouted one person. "You want us to trade with Labor Land. Well, we just found out that wages in Labor Land are much lower than the wages we are paid here in Capital Land. We'll never compete with them. Their costs are so low, they'll charge a lower price for everything. They'll make and sell everything, and we'll make and sell nothing."

"Yeah!" screamed the audience. "Our factories will all shut down and we'll all be out of work!"

"How ironic," Ricardo told the audience. "That's exactly what the Labor Landers said. But they predicted disaster from free trade because we're a productive powerhouse—more productive in making everything than they are. But now you are also predicting disaster from free trade because their wages and costs are much lower. Fortunately, your fear is as baseless as theirs."

"Why?" shouted the audience.

"Let me explain," said Ricardo calmly. "Suppose for a moment you can't sell anything because your wages, costs, and prices are higher, and you want to buy everything from Labor Land because its wages, costs, and prices are lower. With everyone trying to buy goods made in Labor Land, its wages, costs, and prices will rise. With no one trying to buy goods made in Capital Land, our wages, costs, and prices will fall. This will all happen very quickly. Soon, one of our goods will be cheaper, and people in Labor Land will want to buy it. That good will be hitech, and you will specialize in its production, sell it to Labor Landers, and use your earnings to import lotechs. You'll do very well and be better off than you were before trade

began. Now, would you like me to explain how I know you will succeed at hitechs rather than lotechs?"

"No," said the talk show host, "that won't be necessary." No one in the audience was screaming anymore. In fact, half were yawning, and half had fallen asleep.

"But I've got a smashing numerical example," Ricardo said with enthusiasm.

The talk show host saw a frantic cut sign from the control room. He led Ricardo by the hand off the stage. Later, he learned that his show's ratings had plunged drastically during the five minutes that Ricardo had been speaking. True, many TVs remained tuned to his station, but only because the viewers had fallen asleep.

"Somehow I feel this has all happened before," said Ricardo wistfully.

Valid Concerns

"Ricardo, we'd like to have a word with you."

"Who are you four gentlemen?"

"We're Heckscher, Ohlin, Stolper, and Samuelson."

"What can I do for you?" asked Ricardo.

"Please don't misunderstand us," said Samuelson. "We have the greatest admiration for your theory of comparative advantage and your demonstration that free trade benefits both lands even when one land is more productive in making every good. As economists, we find your reasoning and numerical examples to be as exciting as life on this earth ever gets."

"Thank you, kindred spirits," said Ricardo.

"However," said Samuelson, "there is one shortcoming from free trade that we economists should confess."

"Please explain," said Ricardo.

"It's important to recognize," said Heckscher and Ohlin in unison, "that each good in each land is made with two factors of production: labor and capital. We've been able to show that each country will successfully export the good that uses a lot of its abundant factor. So Capital Land will export hitechs because hitechs are produced with a lot of capital, and Labor Land will export lotechs because lotechs are produced with a lot of labor."

"Your analysis with two factors of production, labor and capital, is

simply fascinating," cried Ricardo. "I used only labor in my analysis."

"We take it a step further," said Stolper and Samuelson in unison. "We can show that free trade helps a land's abundant factor and harms a land's scarce factor. In other words, free trade will raise the real wage of low-skilled workers in Labor Land where such labor is abundant, but will reduce the real wage of low-skilled workers in Capital Land where such labor is scarce. The reason is simple. Free trade causes Labor Land to expand production of lotechs, which uses a lot of low-skilled labor, so wages get bid up. But free trade causes Capital Land to shrink production of lotechs, so wages of low-skilled workers get forced down."

"But you do find that each land gains from free trade, don't you?" asked Ricardo.

"Yes, we do," replied Samuelson. "The gain to everyone else in Capital Land is much greater than the loss to low-skilled workers through the fall in their wages. To be precise, we show that if everyone who gains from trade compensated those who lose, then everyone in the land would be better off."

"Excellent," said Ricardo.

"However," said Samuelson, "we economists should surely advocate not only free trade, but compensation to those who lose from it."

"How should the compensation be implemented?" asked Ricardo.

"In two ways," replied Samuelson. "First, taxpayers should help finance the retraining of workers who lose jobs when free trade causes some industries to contract. And second, in Capital Land, we must explain that because free trade may well harm low-skilled workers, it is important to have a progressive tax system, an earned income credit, and social insurance to compensate these workers so that free trade will in fact benefit everyone."

Two citizens who had been listening now spoke up. "I have another concern," said Labortas. "Labor Land has much weaker labor standards than we do. For example, they permit child labor, poor working conditions, and very long work days. As a result, they out compete us in goods involving such labor. Even worse, our industries lobby to weaken our labor standards, complaining that there should be a level playing field."

"Why, your concern is similar to mine," said Environmentas. "Labor Land has much weaker pollution standards than we do. As a re-

sult, they out compete us in goods involving pollution. Even worse, our industries lobby to weaken our pollution standards, complaining that there should be a level playing field."

"You've both raised valid concerns," said Econo, one of the brightest young economists in Capital Land. "We economists must show some common sense. In our standard models, people get satisfaction only from consumption of goods. But our people actually care about whether production involves the exploitation of labor or the degradation of the environment. Once we recognize that these are legitimate preferences of our people, sensible economists should have an open mind about proposals to restrict trade when exploitation of labor or environmental degradation is severe."

"But we must be careful," warned Ricardo. "When a land becomes open to free trade, the industries in which it has a comparative disadvantage are forced to contract by foreign competition. Most people in both lands benefit from trade, but naturally, workers and managers in the contracting industries are unhappy. They are likely to accuse the other land of extreme exploitation or pollution, whether valid or not, to get restrictions on trade and protect their position."

"Very true, Ricardo," said Econo. "We must use our common sense and strike a balance. Labor Land is less productive than we are, and therefore has a lower wage and a lower standard of living. Labor Landers may feel they can't afford to match our labor and environmental standards at this stage of their development. We should, therefore, not insist on such a matching in order to trade with Labor Land. At the same time, we should insist that Labor Land avoid extreme labor exploitation and environmental degradation before we are willing to trade." Econo's advice was followed. In the trade agreement signed by the governments of Labor Land and Capital Land, Labor Land agreed to improve its working conditions, reduce its work day, eliminate child labor, and reduce its pollution. Capital Land did not insist that wages, labor conditions, or environmental quality match its own. Capital Land enacted programs to help its low-skilled workers cope with the effects of trade.

And so, our tale of two islands comes to a happy end. Having overcome their baseless fears, and having responded to valid concerns, the two islands moved boldly forward, trading freely, and benefiting mutually.

III

Macroeconomics

5

Recession

Recession is no time to be looking for your first full-time job. *Recession* means that the total output (gross domestic product, or GDP) of the economy falls, workers are laid off and become unemployed, pay raises are smaller for those who remain employed, company revenues and profits fall, the stock market usually falls, and most important of all (I'm sure you will agree), college graduates seeking their first full-time jobs have difficulty finding them.

In a recession, the unemployment rate rises above normal. The *unemployment rate* is defined as the percentage of people in the labor force who are without work. To be counted in the labor force, you must be actively seeking work. In the United States, the normal unemployment rate is roughly 5.5 percent: for every 1,000 people actively seeking work, 55 of them are without work (and 945 are employed). In the recession of the early 1980s, the unemployment rate rose above 10 percent; in the recession of the early 1990s, above 7 percent; and in the recession of the early 2000s, above 6 percent. Even in a healthy economy, there is always some unemployment, because people sometimes decide to quit one job before finding another, employers sometimes fire workers who are performing badly, people who are out of the labor force (most full-time college students) often require some time to find a job once they decide to actively seek employment, some of the unemployed lack the skills sought

by employers trying to fill vacancies, and some of the unemployed live in one geographic area while there are job vacancies located in another geographic area. In chapter 6, we'll mention some policies that can reduce the normal unemployment rate of the economy. But in this chapter, we concentrate on what to do if the unemployment rate rises above normal.

Some recessions cure themselves quickly, just as some illnesses do. But some recessions drag on for a long time, and some get worse—the recession becomes a *depression*. Just as some illnesses require proper medicine to help the patient recover, so some recessions require proper economic medicine to help the economy recover. This crucial lesson is taught by John Maynard Keynes in his book, *The General Theory of Employment, Interest, and Money* (1936). What kind of economic medicine? The first medicine is a temporary cut in interest rates to stimulate borrowing and spending (demand). Who administers it? Our central bank, the Federal Reserve ("the Fed"). This medicine is called *expansionary monetary policy*. If it doesn't do the trick, there is a second medicine: a temporary cut in taxes and/or temporary increase in cash transfers to households so that they can afford to spend (demand) more. Who administers it? Congress and the president. This medicine is called *expansionary fiscal policy*. Some economists prescribe giving the patient both medicines immediately, rather than in sequence. Both medicines must be stopped as soon as the patient recovers. Use beyond that could result in a harmful overdose, as we will see.

But before we go any further, we need to see what determines the total output (GDP) of the economy. In chapter 1, in a section entitled, "Accounting: Output (GDP), Income, Consumption, Investment, and Saving," we introduced the concept of GDP. Go back and re-read this section right now. . . . Welcome back. In chapter 2, we saw that demand and supply determined the amount of gasoline and other individual products that are actually produced. The same is true for the whole economy. It is time to apply demand and supply to the whole economy to understand what causes a recession, and how proper economic medicine can help cure it.

Demand

It is useful to divide the demanders—the buyers—into four groups. First, there are households who demand consumption (C) goods.

Second, there are business firms that demand investment (I) goods (such as machinery and new technology). Third, there is government (G) that demands goods such as planes and tanks and computers. Fourth, there are foreigners who demand our exports (X). So *aggregate* (total) demand (D) equals C + I + G + NX (we'll explain the N in NX in a moment). Let's make a few comments on each component of D.

Consumption demand (C) varies directly with household income—more precisely, with household "disposable" income, which is income after taxes but including cash transfers; this is the amount of income the household has at its disposal. Hence, C is stimulated by a temporary cut in taxes or increase in cash transfers. C is also stimulated when the Fed induces banks to cut interest rates so that households can borrow at cheaper rates to buy durable goods such as automobiles and home appliances. Of course, C also depends on household optimism and pessimism—for example, if households turn pessimistic, fearing a recession that might make them unemployed, they are inclined to prepare for it by saving more and consuming less.

Investment demand (I) varies with business managers' expectations of future sales. When managers are optimistic, they order new machines and technology, raising I; if they fear a recession, they cut I. I is stimulated when the Fed induces banks to cut interest rates so that businesses can borrow at cheaper rates to buy machinery.

Government demand (G) depends on authorization by Congress and the president. Note an important point. G is government demand for goods—like planes, tanks, and computers—and for labor services—like the labor of military personnel. G is *not* government cash transfers to households or businesses, like Social Security benefit checks. Cash transfers to households cause C to increase; they are not part of G.

NX stands for net exports. Why the "net?" D is aggregate demand for goods and services produced in our own country. While we must include in D foreign demand for our exports (X), we must subtract demand by our households, businesses, and government for goods produced in other countries—our imports (M). Instead of subtracting demand for imports from C, I, and G, it is more convenient to subtract M from X, and call it NX, net exports.

Supply

One point about supply deserves emphasis: the height of the supply curve reflects the cost of production. If the cost of production increases, the supply curve shifts up, and if the cost decreases, the supply curve shifts down. For example, in the mid-1970s, the Persian Gulf countries sharply raised the world price of oil. This increased the cost of production in the United States, and shifted up the U.S. supply curve. By contrast, in the mid-1990s, advances in computer and information technology reduced the cost of production in the United States and shifted down the U.S. supply curve.

Demand and Supply in Recent Decades

Shifts in demand or supply change output and price. A rise in price is called *inflation*. A fall in output is called *recession*. When output falls, employment falls, so unemployment rises. For example, suppose 1,000 people want to work. In a normal economy, 945 obtain jobs and 55 are unemployed, so the *unemployment rate* is 5.5 percent (55/1,000). If something makes aggregate demand fall, then output falls, so firms lay off some employees. To illustrate, if firms lay off 20 workers, employment falls to 25 and unemployment rises to 75, so the unemployment rate rises to 7.5 percent (75/1,000).

Figure 5.1 shows a D/S diagram for each of the past five decades in the United States. In the late 1960s, the Vietnam War caused an increase in military spending, raised G, and shifted the D curve to the right, causing an increase in output (and employment), but also an increase in prices ("inflation"). In 1974, and again in 1979, the world price of oil rose sharply, raising the cost of production in the United States, shifting up the S curve, causing inflation and a fall in output (and employment)—a recession. In the early 1980s, to slow the inflation the Federal Reserve raised interest rates, thereby reducing borrowing and spending (demand) by households and businesses, shifting the D curve left, causing a reduction in inflation, but also a fall in output (and employment)—a recession. In the mid 1990s, advances in computer and information technology reduced the cost of production, shifting down the S curve, causing an increase in output (and employment) and a fall in inflation. Early in the 2000

Figure 5.1   **Demand and Supply in Five Decades**

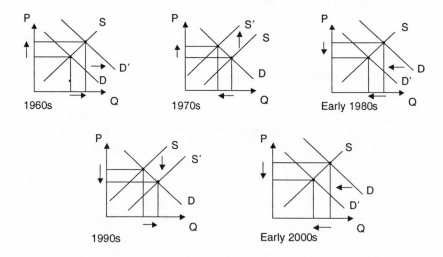

decade, the stock market fell sharply, and pessimism caused business firms to cut investment, shifting the D curve to the left. The result was a recession.

The Fed, Open Market Operations, and Interest Rates

To fight a recession, the Fed induces banks to cut interest rates they charge households and businesses, thereby encouraging borrowing and spending (demand), shifting the D curve to the right. But exactly how does the Fed induce banks to cut interest rates? By injecting more money into the banking system. How? By *open market operations*, or buying government bonds from the public (households, businesses, or local governments) in the same "open market" where the public buys and sells government bonds. The sellers of bonds deposit the checks in their banks, and these banks in turn request and obtain an amount of cash from the Fed equal to the amount on its checks. In response to the infusion of cash "reserves," banks try to increase lending—that's how banks make a profit. To attract borrowers for their excess cash, banks compete by reducing interest rates.

One way to grasp how the Fed achieves a cut in interest rates is by envisioning a demand/supply diagram for loanable funds. On the

horizontal axis is the quantity of loanable funds, and on the vertical axis is the interest rate. The suppliers of loanable funds are lenders, such as banks. The demanders of loanable funds are borrowers, such as business firms. When the Fed injects funds into the banks through open market operations, the banks increase their supply of loanable funds, so the supply curve shifts to the right, thereby reducing the interest rate.

But where does the Fed get the cash to buy bonds in the open market? The Fed is permitted to issue new cash. Take out your wallet and look at the green paper ($1, $5, $10, $20, etc); at the top it says "federal reserve notes." Isn't it risky to give any institution the power to issue new cash? It certainly would be risky to give it to Congress, where politicians may be tempted to pay for popular spending programs by printing new money instead of raising taxes. But, as we will see in a moment, the Fed is better insulated from political pressures and temptations. Usually it refrains from issuing too much new money.

But why must the Fed use its new cash to buy government bonds? The Fed simply wants to inject new cash into the banks, so buying anything from the public would do the trick. Whatever the Fed buys, the seller will deposit the check at his bank, and the bank will request and obtain an equal amount of cash from the Fed. Yet, in practice, the Fed buys only government bonds. Why? Because by law that is all the Fed is permitted to buy. Why did Congress restrict the Fed to government bonds? Because it did not want the Fed selecting bonds, stocks, or products of particular companies; it was thought that such a selection process might be vulnerable to corruption—companies or politicians might try to bribe Fed members. So to be safe, the Fed buys government bonds, and this gets the cash into the banks and reduces interest rates.

So who decides how much cash the Fed should inject through open market operations? The Federal Open Market Committee (FOMC) which meets monthly. Who's on the FOMC? First, there are the 7 members of the board of governors of the Federal Reserve System. Second, there are the 12 presidents of the 12 regional Federal Reserve banks; at any meeting of the FOMC, all participate in discussion, but only 5 vote (they rotate from meeting to meeting). Each member of the board of governors is appointed by the president for a

fourteen-year term, and the president selects one member to be chairman of the Fed for a four-year term (the Fed chairman can be reappointed). The current (2003) chairman of the Fed is Alan Greenspan, who has held the position since 1987; the preceding Fed chairman was Paul Volcker (1979–87). The Fed is certainly subject to popular and political pressures, but it is clearly much more insulated than Congress, though not quite as insulated as the Supreme Court, whose members serve for life rather than 14 years.

Suppose that even after the economy is back to normal GDP, the Fed keeps injecting cash into the banks through open market operations so that interest rates stay low and borrowing and spending keep rising. Then the D curve keeps shifting to the right. The result is a rise in prices—inflation. In chapter 6, we will see why this would be harmful. Thus, it is true that continuing to inject cash into the banking system after GDP has returned to normal would lead to harmful inflation. Fortunately, the Fed is insulated from political and popular pressures, and usually refuses to overdose the economy with money.

Automatic Fiscal Policy

But cutting interest rates may not be enough to successfully combat a severe recession. Sometimes, whether the Fed wants to admit it or not, it needs help. If the D curve has shifted far to the left generating a severe recession, cutting interest rates may not be able to push the D curve all the way back to the right to its normal position.

Help must come from Congress and the president through a temporary cut in taxes and/or increase in cash transfers. Sometimes they act quickly—for example, they promptly enacted "tax rebates" in both the 1975 and 2001 recessions. But often they delay and deadlock, and act either too late or not at all.

What can be done about the delay? The answer is to have Congress pre-enact a temporary tax cut and/or cash transfer that would be automatically triggered as soon as the economy falls into recession—the deeper the recession, the larger would be the tax cut or transfer. Such an automatic fiscal policy would help the Fed combat severe recessions.

Suppose Congress enacts the following law. Whenever the U.S. Department of Commerce reports that last quarter's GDP was more

than, say, 2 percent below normal, a cash transfer ("rebate")—a check from the U.S. Treasury—would be immediately mailed to all households; the larger the GDP gap, the larger would be the amount of the check. In June 2001, Congress passed a tax law that instructed the U.S. Treasury to mail a $600 check to most households to combat the 2001 recession. The checks were mailed in July, August, and September. But in the fall of 2001, partisan differences kept Congress from authorizing a second round of checks, despite the continuing weakness of the economy. A permanent law would assure that checks are automatically sent out each quarter until the economy is strongly recovering.

But where does Congress get the cash to back its checks to households? After all, households will quickly deposit the checks in their banks, and their banks will promptly request cash from the U.S. Treasury equal to the amount on the Treasury checks. Should Congress get the cash by raising taxes? No, because that would defeat the whole purpose of the policy—to get more cash into people's bank accounts so they can afford to spend (demand) more. To achieve its purpose of fighting the recession, Congress must run a temporary *budget deficit*—it must spend more than it raises in taxes. How about just printing new cash? Wisely, Congress is prohibited from simply printing new cash whenever it claims it needs it. Historically, letting politicians print new money whenever they claim they need it has led to an overdose of cash in the economy which results in a harmful rapid rise in prices because the D curve keeps shifting right (up).

So Congress must borrow the cash. From whom? Congress is prohibited from borrowing directly from the Federal Reserve, but it can do it indirectly. Here's how. Congress authorizes the Treasury to sell ("issue") bonds to the public. The public buys the bonds and the Treasury gets the cash it needs to back its checks to households. Then the Fed, through open market operations, buys an equal amount of bonds from the public. So what has happened? The public ends up with the same amount of bonds and cash it started with, the Treasury ends up with more cash, and the Fed with more government bonds. Indirectly, the Treasury has borrowed cash from the Fed—the public has been the intermediary.

As long as the Fed approves of the expansionary fiscal policy, and

buys an equal amount of bonds, the fiscal policy also results in an injection of more money into the economy. The Treasury goes into debt, not to the public, but to the Fed which ends up with the Treasury's bonds. And the Fed is a nice creditor. If the Fed decides that the Treasury has issued the bonds for a good reason—to fight a recession—then the Fed can say to the Treasury, "Don't worry about paying interest or even the principal—you borrowed for a good purpose."

Note the importance of the separation of powers between the Treasury, which represents Congress, and the Fed. Suppose in another situation the Fed disapproves of the Treasury's borrowing, because there is no recession to fight, but Congress simply wants to spend money on popular programs but is unwilling to raise taxes. Then the Fed refuses to buy the bonds through open market operations, and the public keeps the bonds it buys from the Treasury. The public is not a nice creditor: it insists that the Treasury pay interest and principal when each bond comes due. So a tough Fed provides some discipline for Congress.

An Always-Balanced Budget Rule Could Turn a Recession into a Depression

So how should Congress set its budget—spending and taxes—each year? A government, like a household or business, should avoid borrowing excessively year after year, because it will be burdened with interest payments, and eventually with repaying its debt. But should the government try to balance its budget every year? Many citizens say yes, but most economists say no. Economists believe that it would be dangerous to try to balance the budget every year. We just saw that in a recession it is better for Congress to raise its spending—send out cash transfers to households—and borrow the money rather than raise taxes, so that a temporary budget deficit is exactly what is needed in a recession.

To many citizens, the remedy for government deficits seems simple: Require a balanced budget every year. According to this view, the planned budget should always aim at balance, based on the best available forecast. Once the fiscal year has begun, if a deficit begins to emerge, a prompt cut in spending or increase in taxes to restore balance should be required. I'll call this an "always-balanced budget

rule." Another name might be "the no ifs, ands, or buts balanced budget rule" or "the no excuses balanced budget rule." The rule is so simple. It's a shame it suffers from a fatal defect: It would destabilize the economy. Why?

Suppose the economy is operating at a normal level of output when the budget for next year is planned. Technicians advise Congress on where to set tax and spending rates so that if the economy remains at a normal unemployment rate, the budget will be balanced.

How do the technicians arrive at their conclusion? They estimate how much tax revenue will be raised—given the statutory tax rates—if the economy is normal and national income is normal; they then compare this revenue to estimated expenditure. Make note of this: If national income turns out to be below normal, then tax revenue will be lower than the technicians' estimate.

As the fiscal year begins, suppose the economy falls into recession. National income falls below normal and, automatically, tax revenue falls below the level that planners estimated. Hence, the budget moves into deficit.

If you are inclined to blame Congress for everything, please note that this particular deficit is not the fault of Congress. The source of this deficit is the unexpected recession. But this is a "no excuses" balanced budget rule. Even though Congress did not cause the deficit, it must act promptly to eliminate it. Under the always-balanced rule, Congress must promptly cut spending or raise tax rates to eliminate the deficit brought on by recession.

What happens to the economy, already in recession, when Congress cuts spending or raises taxes? Suppose Congress cuts government purchases. For example, it cuts the purchase of planes for the military and computers for government offices. Then the producers of planes and computers suffer a fall in orders and hence cut production. Their employees earn less income and in turn cut their consumer spending. The recession deepens.

Or suppose Congress cuts cash transfers. For example, it cuts spending on welfare, food stamps, and college financial aid. Then the recipients of these transfers have less to spend, and producers observe a decline in demand. They cut production. Their employees earn less income and in turn cut their consumer spending. Once again, the recession deepens.

Finally, suppose Congress raises tax rates. Taxpayers have less after-tax income, cut their spending, and the result is the same: an intensification of the recession. Thus, no matter how Congress tries to eliminate the recession-induced deficit, it makes the recession worse. For decades, economics textbooks have emphasized that an always-balanced budget rule risks turning a recession into a depression.

NUBAR

Does this mean that every balanced budget rule is dangerous? No. There is a safe balanced budget rule that permits a temporary deficit in a recession. I call it NUBAR—my own acronym—which stands for "normal unemployment balanced budget rule," and which reads as follows: "Congress shall enact a *planned* budget for the coming fiscal year that technicians *estimate* will be balanced *if next year's unemployment rate is normal (the average of the preceding decade)*." The technicians should probably be the employees of the Congressional Budget Office (CBO), who already perform similar tasks for Congress.

Note two crucial features of NUBAR: First, it applies to this year's *planned* budget for next year, not this year's actual spending and revenue. Second, it is *not* based on a *forecast* of next year's economy. Technicians are instructed to estimate spending and tax revenue on the assumption that next year's unemployment rate will be normal, whether or not they forecast a normal unemployment rate for next year.

Assume the technicians are accurate in their estimates about what spending and tax revenue will be if next year's unemployment rate is normal. If next year's unemployment rate turns out to be normal, the budget will be balanced; if next year's unemployment rate turns out to be above normal (national output, income, and tax revenue below normal), the budget will run a deficit; and if next year's unemployment rate turns out to be below normal (national output, income, and tax revenue above normal), the budget will run a surplus. On average, but not in every year, the budget will be balanced. Even if the technicians make errors in particular years, as long as the errors are unbiased, on average the budget will be approximately balanced.

With NUBAR, what happens if the economy falls into recession? Automatically, tax revenue falls and a budget deficit results. Under NUBAR, Congress is not required to immediately raise taxes or cut spending in the recession because NUBAR applies to the *planned* budget, not current spending and revenue.

What about the budget that is planned in the middle of recession for the coming fiscal year? Under NUBAR, Congress must set spending and tax rates so that technicians estimate that the planned budget will be balanced *if and only if* the economy returns to a normal unemployment rate next year. Suppose someone objects, "But this year's recession is so deep that output, income, and tax revenue will still be below normal next year, so next year the actual budget will show a deficit."

A NUBAR advocate should reply, "That's exactly what we want. We don't want to achieve a balanced budget next year if the economy is still in recession. Why not? Because to achieve a balanced budget next year in recession, this year we would have to plan higher tax rates, and lower spending rates, than NUBAR requires. But these higher tax rates and lower spending rates would reduce total spending in the economy next year, making next year's recession worse. We expect NUBAR to result in an actual deficit in recession, and an actual surplus in a boom. That's what we want."

NUBAR is not a novel proposal. For many years, economists have recommended that Congress try to balance the planned budget on the assumption that next year there is *full* employment. The only difference is that NUBAR requires the technicians to assume that there will be a normal unemployment rate (the average of the preceding decade), instead of assuming that there will be full employment. Thus, the basic strategy of NUBAR has long been advocated by economists.

There are, however, two differences between NUBAR and the full employment balanced budget rule. First, NUBAR avoids a debate about what is "full employment." Instead, NUBAR uses the normal unemployment rate as its benchmark. The normal unemployment rate is defined as the average of the preceding decade. "Normal" is not necessarily "optimal" or "full."

Second, NUBAR will achieve an average budget deficit, over a decade, that is close to zero, because it is based on a realistic unem-

ployment rate—the actual average of the preceding decade. A full employment budget rule will achieve an average budget deficit, over a decade, that is much greater than zero. The reason is that, in practice, Congress will undoubtedly define "full employment" more ambitiously than normal employment.

For example, under NUBAR, if the average unemployment over the preceding decade has been 5.5 percent, then NUBAR will require a planned budget that is estimated to be balanced on the assumption that the economy's unemployment rate will be 5.5 percent. We do not claim 5.5 percent is "optimal" or "full"; it is simply realistic, "normal." A full employment balanced budget rule might require a planned budget that is balanced on the assumption that the economy's unemployment rate will be, say, 4.5 percent, or even 4 percent. But when the economy's unemployment rate turns out, on average, to be closer to 5.5 percent, this planned budget will result in a deficit. Thus, in practice, NUBAR will on average balance the budget, while a full employment balanced budget rule will not.

You might be thinking, "True, NUBAR does not force Congress to take action that makes a recession worse. But does it fight a recession?" NUBAR itself does not; NUBAR itself is neutral. Fighting a recession is the job of the automatic fiscal policy described above— the temporary tax cut and/or cash transfer that is automatically triggered by a fall in GDP. Although NUBAR itself doesn't fight a recession, it permits the automatic fiscal policy to fight it. Why? Because this automatic triggering in a recession does not violate NUBAR. NUBAR requires only that the planned budget be estimated to be balanced if the economy is normal.

But why make Congress adhere to any balanced budget rule, even a safe one like NUBAR? Most (not all) economists believe that it is a good thing to have a properly constructed balanced budget rule for Congress because legislators need fiscal discipline to avoid borrowing excessively year after year, which would eventually burden the government and, hence, future taxpayers with huge interest payments. Legislators will often be tempted to propose an increase in spending, or a cut in taxes, because both actions are popular. But a balanced budget rule imposes some discipline. With a rule, legislators must face the fact that raising spending will require raising taxes, and cutting taxes will require cutting spending. They will think twice before

proposing an increase in spending, because opponents will point out that their proposal requires an increase in taxes. And they will think twice before proposing a cut in taxes, because opponents will point out that their proposal requires a cut in government spending. The simplest way to impose this discipline is to have Congress adhere to a balanced budget rule. A NUBAR statute would impose some discipline on Congress, but at the same time would be safe in a recession. NUBAR plus automatic fiscal policy would both provide discipline in normal times and stimulus during a recession.

6

Inflation

What is inflation? Inflation is a rise in most prices—a rise in the *average price level* of the economy. What causes it? A shift up of the D or S curve of the economy. As we saw in chapter 5, in the 1960s a shift up (right) of the D curve due to Vietnam War spending caused prices to rise; and in the 1970s, a shift up (left) of the S curve due to the OPEC oil price increases caused prices to rise. When the cause of inflation is a shift up of the D curve, the inflation is accompanied by a boom—a rise in the gross domestic product (GDP) and a fall in unemployment. When the cause of inflation is a shift up of the S curve, the inflation is accompanied by a recession—a fall in GDP and a rise in unemployment. Oversimplifying, we can say that the shift up of demand in the 1960s caused inflation to rise from 0 percent to 5 percent per year, and the shift up of supply in the 1970s caused inflation to rise from 5 percent to 10 percent per year.

History shows that the worst episodes of inflation have virtually always been caused by a shooting up (right) of the D curve due to huge government budget deficits financed by printing money. Huge inflation—"hyperinflation"—has virtually always been caused by the government pouring huge amounts of money into the economy to pay for the excess of its spending over its tax revenue. It happened in the U.S. South (the Confederacy) during the Civil War. It happened in Germany during the 1920s. It happened in several Latin American

countries in recent decades. The prevention of hyperinflation is therefore straightforward: Unless the economy is facing a severe recession due to a drastic drop (shift left) of the D curve, don't run huge budget deficits financed by printing money.

How Harmful Is Inflation?

A hyperinflation is extremely harmful. People become obsessed with trying to spend money before its "value"—the goods and services it can purchase—declines. Workers want to shop immediately after getting a paycheck. They ask employers to pay them more frequently and give them frequent shopping breaks so they can purchase goods immediately. People spend more time shopping and less time working, causing output to decline. In fact, workers ask employers to pay them with goods rather than money. Participants in the economy regress to barter, where people try to exchange goods for goods, rather than goods for money. But barter is enormously time consuming—each seller of labor or particular goods must find a buyer who has the particular goods the seller wants. Hence, money—a powerful invention to avoid wasting time—is abandoned, and a huge amount of time is spent on bartering instead of producing goods and services.

By contrast, a moderate inflation—say 10 percent per year (the U.S. inflation rate at the end of the 1970s)—does not cause people to abandon money for barter. So how harmful is a moderate inflation? At the end of the 1970s in the United States, public opinion polls showed that 10 percent inflation was highly unpopular. Why? Most people assumed that if inflation were brought down, their wage increase would be unaffected. If this were really true, they would obviously benefit from a reduction in inflation—they would be able to buy more goods and services.

But this widespread assumption was (and is) erroneous. To see the error in an instant, put yourself in the position of a business manager. To stay in business, you must set your price high enough to cover your cost. If unit cost (cost per unit of output) increases, you must raise your price to cover the increase. The most important cost in most businesses is labor cost. In 1980, "wage and salary increases averaged 11 percent a year. Since output per worker—"labor productivity"—was increasing (on average) only 1 percent per year,

unit cost was (on average) rising 10 percent per year. To cover this cost increase, business managers were (on average) raising prices 10 percent per year. It was obvious to business managers that they could afford to reduce their price increase—hence, reduce inflation— *only* if they were able to reduce their wage increase. Thus, the public's expectation that inflation could be reduced without reducing their wage increase was clearly in error. More specifically, business managers could afford to reduce their price increase from 10 percent to, say, 0 percent, only if they could reduce their wage increase from 11 percent to 1 percent. But then workers would be no better off. The reason should be obvious. If (labor) productivity increases only 1 percent per year, then workers (on average) can only buy 1 percent more goods and services each year because you can't buy what you don't produce.

But even if a worker wouldn't benefit from a reduction in inflation, wouldn't a saver benefit? Yes, but only because the United States overtaxes savers when there is inflation. To see this, first assume there is no income tax. Consider a saver who put $100 in the bank. In the late 1970s, interest rates were much higher than today, so suppose the saver could get 12 percent interest. But with inflation at 10 percent, at year's end the saver would need $110 to buy the same goods and services that could have been bought for $100 at the beginning of the year. The saver would have $112, so the benefit—the "real" interest rate (after adjusting for inflation)—was approximately 2 percent. The saver would obviously benefit if the interest rate would stay at 12 percent when inflation was reduced. But it turns out that interest rates generally move up and down (eventually, not immediately) roughly point for point with inflation. So if inflation is cut from 10 percent to, say, 0 percent, eventually (not immediately) the interest rate will probably fall from 12 percent to 2 percent, so once again, the saver's benefit will still be 2 percent.

But now consider what happens under the U.S. income tax. Suppose this saver is in a 25 percent tax bracket. With inflation at 10 percent and the interest rate at 12 percent, the saver would be taxed 25 percent of the $12 of interest, or $3, leaving $9 of interest after tax, so the saver would have $109 at the end of the year. But prices are 10 percent higher, so the saver would lose 1 percent. By contrast, with 0 percent inflation and the interest rate at 2 percent, the saver

would be taxed $0.50 on the $2 of interest, leaving $1.50, so the saver would have $101.50, and would therefore gain 1.5 percent. Most economists believe the income tax rules should be changed so that only interest above inflation is subject to tax. But until that happens, savers in the United States are better off when inflation is lower.

There is, however, a different kind of benefit from reducing inflation: psychological. With 10 percent inflation, many workers and savers feel unhappy and frustrated because they mistakenly believe they are being "robbed" 10 percent. It's not true, but history suggests that the majority of people feel this way about inflation. This unhappiness is real even if it is based on error. A reduction in inflation would reduce this unhappiness.

There are some benefits of disinflation (reducing inflation) that are not based on error. People holding cash (in their pockets and homes) do not receive any interest, so a reduction in inflation would better preserve the value—the purchasing power—of their cash. Inflation harms the holders of cash ("currency"), so disinflation would benefit them. Also, retirees receiving a fixed monthly pension benefit are hurt by inflation and helped by disinflation. And, once again, savers are better off due to the income tax.

So does this mean our goal should be zero inflation? Probably not. It would probably be best to have 2 percent inflation. Why? Two reasons. First, interest rates would usually be 2 percent higher (say, 4 percent instead of 2 percent). Now suppose a recession hits. The Fed would be able to cut interest rates 4 percentage points, instead of only 2, to fight the recession. Thus, if inflation and interest rates are too low, the Fed has less power to combat recessions.

Second, the average wage increase would be 2 percent higher (say, 3 percent instead of 1 percent). In a dynamic economy, demand is always rising for some products and falling for others. When demand for a product falls, business managers can't afford the same wage increase. Suppose cutting the wage increase 2 percent below average is required. If the average is 3 percent, the firm must cut the wage increase to 1 percent. But if the average is 1 percent, the firm must cut the wage increase to -1 percent—it must literally pay fewer dollars this year than last. But experience shows that there is a loss in morale and rise in anger among workers who are given a cut in their dollar wage. Rather than do this, employers may simply lay off some

workers, rather than cut everyone's wage. Hence, unemployment may be higher in a 0 percent inflation economy than in a 2 percent inflation economy. So from this point on we'll assume that the target for inflation should be 2 percent.

Unfortunately, once inflation is 10 percent, we cannot simply wave a magic wand and make it 2 percent. It must be brought down, and the process of bringing it down involves social costs. To see this, let's look at how the 10 percent inflation at the end of the 1970s was brought down in the early 1980s.

Who Dunit?

Oversimplifying, we can say that a shift up of demand in the 1960s from the Vietnam War raised inflation from 0 percent to 5 percent, and the shift up of supply in the 1970s from oil price increases raised inflation from 5 percent to 10 percent. In 1979, with the inflation rate near 10 percent, President Carter appointed Paul Volcker to be chairman of the central bank of the United States—the Federal Reserve. Volcker and his colleagues resolved to do what was necessary to bring down inflation. They applied a "tight money" policy to the economy, long enough and hard enough for the resulting severe recession to bring down inflation.

Let's go through that more slowly. I'll explain what "tight money" means in a minute. But first, why does a deep recession bring down inflation? A deep recession means layoffs. Many workers become alarmed that they will lose their jobs. So instead of being aggressive at the bargaining table, unions are willing to make concessions to save jobs. Not only that, the recession also means poor profits, and employers simply can't afford large wage increases. So a deep recession slowly brings down wage increases.

Business firms set price increases to cover cost increases. When wage increases get smaller, business firms can afford to reduce price increases. And competition forces them to. Also, in a recession, market demand won't support customary price increases. So price increases get smaller. Hence, inflation declines.

The members of the Fed knew that disinflation would not be a very pleasant process. I'm sure they undertook their assignment with great regret. But there is no escaping the central point. The Fed can

bring down inflation only by generating a recession. And that's precisely what it did at the beginning of the 1980s.

How did the Fed do it? By making money "tight." What does this mean? The Fed raised interest rates high enough to discourage many households and businesses from borrowing. These households and businesses were forced to cut their spending on goods and services. When producers confronted the fall in demand, they had no choice but to cut production and lay off workers.

But how does the Fed raise interest rates? Every month, unbeknownst to most of the public, the Federal Reserve's Open Market Committee—the FOMC—meets in Washington. The FOMC consists of the Fed chairman (currently Alan Greenspan, who replaced Paul Volcker), six other board members who have been appointed by the president (each term is fourteen years), and the twelve Fed regional bank presidents. They are advised by an excellent staff of well-trained economists.

Each month the FOMC decides the stance of monetary policy—to tighten or to loosen, that is the monthly question. At the beginning of the 1980s, the FOMC almost always resolved to tighten. It told the Fed manager of *open market operations* in New York to cut down the purchase of government bonds. Let's see how this decision raised interest rates in the economy.

When the Fed buys bonds, the seller—a household, business firm, or governmental unit—deposits the check in its bank. The bank, in turn, obtains cash from the Fed to cover the check. Now you might think that the bank had better hold on to this cash, because, after all, it really belongs to the depositor, who could come in any old time and ask for it. But centuries ago, an early banker experienced the ecstasy of discovering that his depositors would never know if he lent out part of their cash. And by lending it, he could earn interest. So he did. And so have banks ever since.

So whenever a bank enjoys an infusion of cash, it immediately tries to lend out part of it. The more cash it has, the lower the interest rate the bank must offer to get potential borrowers to take all of it. So when the Fed cut down its purchase of government bonds at the beginning of the 1980s, less cash flowed into banks. With less cash, banks could charge a high interest rate and still find enough borrowers for their limited supply of cash.

One way to grasp how the Fed achieves a rise in interest rates is by envisioning a demand/supply diagram for loanable funds. On the horizontal axis is the quantity of loanable funds, and on the vertical axis is the interest rate. The suppliers of loanable funds are lenders, such as banks. The demanders of loanable funds are borrowers, such as business firms. When the Fed reduces its injection of funds into the banks by cutting its open market purchases, the banks experience a decrease in their supply of loanable funds, so the supply curve shifts to the left, thereby raising the interest rate.

By cutting down its purchase of bonds, the Fed raised interest rates throughout the economy. Each month the FOMC would ask: Have we tightened enough? The question was easy to answer. If there was still no recession, then more tightening was necessary. Table 6.1 shows what the Fed was doing: Each month, the FOMC would take the interest rate up, and then wait to see whether the borrowers and spenders would relent. The borrowers and spenders in the economy held out gallantly as the prime rate (the interest rate banks charge their most favored customers) rose to unprecedented heights, as shown in Table 6.1. Finally, in late 1981, the Fed won its battle, and many potential borrowers and spenders surrendered. Confronted with record interest rates, they at last cut down their borrowing and spending. The economy fell off a cliff into the worst recession since the 1930s as the unemployment rate rose sharply in 1982. In the worst month of 1982, the unemployment rate nearly reached 11 percent, and, as shown in Table 6.1, it averaged 9.5 percent for the year.

Table 6.1 also shows that the severe recession succeeded in bringing down wage increases and inflation. With high unemployment, workers were willing to accept smaller wage increases to try to preserve jobs. Smaller wage increases meant smaller cost increases, so firms could afford to set smaller price increases. Just in time, in mid-1982 the Fed relented, let interest rates come down, and prevented the recession from becoming a depression. In fact, a recovery began in 1983 (though it was still not evident in the unemployment rate), and was running at full steam in 1984. Since then, the Fed has usually done a good job of keeping the unemployment rate near normal and the inflation rate low. There was a mild recession in the early 1990s, but the unemployment rate never reached 8 percent. By the mid-1990s, the unemployment rate was

Table 6.1

The Rise and Fall of Inflation (in percent)

	Unemployment rate	Prime interest rate	Wage increase	Inflation
1977	6.9	6.8	8.0	6.4
1978	6.0	9.1	8.9	6.8
1979	5.8	12.7	9.5	8.5
1980	7.0	15.3	10.8	9.7
1981	7.5	18.9	9.7	9.5
1982	9.5	14.9	7.5	6.2
1983	9.5	10.8	4.3	3.2

below 5.5 percent and the inflation rate was about 2 percent. In the early 2000s, the stock market plunged and the economy fell into recession, but the Fed promptly cut interest rates to combat the recession. The recession remained relatively mild, with the unemployment rate rising from 4 percent to 6 percent (of course, it didn't seem mild to college graduates looking for their first job), and inflation stayed below 2 percent.

Should the Fed have generated a recession to bring down inflation in the early 1980s? Most economists agree that inflation had to be brought down. Some would have preferred a milder recession that lasted longer to the sharp, severe recession that occurred. But most concede that it is difficult for the Fed to fine-tune a slowdown of the economy. It is possible that most FOMC members would also have preferred a milder, longer recession.

Could the Fed have used any help? A minority of economists think a wage-price policy could have helped the Fed bring down inflation with less recession by applying some direct pressure to firms to reduce price increases. The majority of economists disagree and think a wage-price policy would have been ineffective. In any case, it's now water under the bridge. Inflation is down, and almost all economists agree that it's up to the Fed to keep it that way.

So let's summarize: The Fed intentionally caused the recession of 1982 and the reduction in inflation that accompanied it. The two went together. Without the recession, there would have been no disinflation. And the Fed generated both.

Miseducation by Politicians

But you would never know "who dunit" if you listened to the rhetoric of the two political conventions of 1984. The Republicans praised President Reagan for conquering inflation but regarded the recession of 1982 as an irrelevant natural disaster, akin to an earthquake. The Democrats blamed the president for causing the worst recession since the 1930s but regarded the disinflation as a mysterious natural blessing, akin to a fortuitous change in the climate. Economists could only sit in front of their TV sets and watch helplessly as millions of Americans were miseducated about recent economic history. How had inflation been brought down? The Republicans cheerfully told the media it was "the president's policies." Which policies, exactly? The media (with a few exceptions) never asked, and the Republicans never said.

Well, that's not quite true. The best of the Republicans never said, because they knew that talking about the Fed's recession was not likely to win votes, so rather than distort any further, they just repeated "the president's policies" and left it at that. Unfortunately, there were other Republicans who went further. They said, "The president's historic tax cut brought down inflation."

Now, the temptation is understandable. After all, to the general public, the president's most famous economic policy was surely the income tax cut he proposed, and persuaded Congress to enact, in 1981. So why not claim that the tax cut did it? Indeed, such a claim would be widely believed for the unfortunate reason that most of the public has not had the opportunity to study any economics.

But few things drive economists wilder than the claim that the tax cut of 1981 brought down inflation. If you took a survey of economists of all political persuasions, there would be near unanimity that it was the tight money policy of the Fed that generated both the severe recession and the resulting reduction in inflation. Don't misunderstand. Many (though not all) conservative economists supported the tax cut of 1981, but not to bring down inflation. For example, Milton Friedman, dean of "monetarist" economists, supported the tax cut because, as a conservative, he wanted to reduce the size of government. But Friedman, who has devoted much of his career to emphasizing the importance of money in the determination of infla-

tion, would be near apoplexy should someone claim that the tax cut of 1981 brought down inflation.

Now what about the Democrats? They claimed that "Reaganomics" had caused the worst recession since the 1930s. But which policies, exactly? They never said. Well, once again, that's not quite true. The best of the Democrats never said, because then it would come out that the Fed did it to reduce inflation. And worse yet, they would have to admit that their own president, Jimmy Carter, had appointed Volcker Fed chairman in 1979, so that if any president was indirectly responsible, it was a Democratic president. So the best of the Democrats just said, "Reaganomics," and let it go at that.

But unfortunately, some Democrats went further. Once again, the temptation is understandable. Since President Reagan's most famous economic policy was his tax cut, these Democrats hinted that the president's tax cut caused the worst recession since the 1930s. Few could get themselves to say that the tax cut caused the recession. But they would say, "The tax cut was a great mistake. Look what happened. We got the worst recession since the 1930s." Now, if there is anything that drives economists wilder than the claim that the tax cut brought down inflation, it is the claim that the tax cut caused the recession. How in God's name can a tax cut cause a recession? A tax cut leaves more cash in people's pockets and raises their spending on goods and services, thereby stimulating production and employment—the opposite of recession.

As for the 1982 recession, there is also a small timing problem. The tax cut was enacted in 1981 to be phased in throughout 1982 and 1983. But the economy plunged into recession before the tax cut took full effect. In fact, the tax cut contributed to the recovery that began in 1983 as a result of the Fed's shifting gears in mid-1982.

So why were the politicians of both parties able to mislead the public? While our nation certainly has some reporters with excellent training in economics, unfortunately many reporters who cover politics have not had the opportunity to study economics, and naturally shy away from asking questions like, "Mr. Republican, exactly how did the tax cut reduce inflation?" or "Mr. Democrat, exactly how did Reaganomics cause the recession?"

It's no accident that it was the Fed, not Congress and the president, that took the unpopular action needed to bring down inflation.

It's not that the Fed is necessarily wiser than Congress and the president, but because it is more insulated from political and popular pressure. First of all, much of the public has never heard of the Fed. How many people understand how the Fed influences the fluctuations of the economy? Second, the members of the FOMC don't run for office every two, four, or six years.

Consider this scene from mid-1982. It's a dinner party in suburban Washington, D.C. But instead of cheer, there is gloom and tension around the table. As one guest empties his wine glass for the sixth time, he whispers to his neighbor that his business is about to go under, ruined by the worst recession since the 1930s.

"That son of a 'bleep' in the White House will never get my vote again. And the whole damned Congress should be drowned in the Potomac," he mutters in anguish. Then he asks his neighbor, "And what do you do?"

"I'm a banker," comes the reply. All around the dinner table, denunciations of the president and Congress can be heard. One man has lost his job, but no one knows it. Several others fear for theirs. Only the banker seems calm.

With good reason. Not only is his job secure, but no one at the dinner party knows that he is no ordinary banker. In fact, he has been a member of the FOMC for several years. His monthly votes have generated the recession that has caused the immediate pain around the table. The resulting disinflation will set the stage for a strong recovery that will last for the rest of the decade. But the pleasure of that recovery is still in the future. Perhaps as he voted, he was strengthened by the knowledge that no one in his social circle would ever know his role or responsibility.

Now, anyone who believes in democracy must have mixed feelings about such a dinner party scene. It is troubling that symmetrical stabilization policy, necessary for the long-run health of the economy, seems to depend on insulation from popular and political pressure. But after all, isn't this the reason we try to insulate the Supreme Court?

So we have two choices. We can exhort Congress and the president to be courageous, and count on them to vote for fiscal medicine when it is medicine we need; or we can face political reality and be thankful that we have another option, to count on the Fed.

The Fed's Target

Once inflation is down, what must the Fed do to keep it low? Experience over the past three decades seems to suggest that if the U.S. economy is run near a particular unemployment rate—perhaps 5.5 percent currently—then the inflation rate usually stays roughly constant. If the unemployment rate is much below this particular value, then inflation usually rises; and if the unemployment rate is much above this particular value, then inflation usually falls. For example, in the late 1960s, the unemployment rate was near 4 percent, and inflation rose. In the 1982 recession, the unemployment rate exceeded 9 percent, and inflation declined, as we saw in Table 6.1.

So if we accept the hypothesis that the economy has a *constant inflation* unemployment rate, or CIRU (pronounced "see-roo"), and if we accept this CIRU as our unemployment rate target, then our two goals are compatible. Incidentally, many economists call the CIRU the *natural rate of unemployment.* I prefer CIRU, to emphasize the point that if the economy is at this unemployment rate, the inflation rate usually stays constant. Other economists call it the NAIRU (nonaccelerating inflation rate of unemployment). But *constant* is clearer than *nonaccelerating*, so I prefer CIRU. At any rate, if you read an article that talks about the natural rate of unemployment or the NAIRU, the author means the same thing that I mean by CIRU.

There is considerable uncertainty about the numerical value of the CIRU. During most of the 1980s, the CIRU appeared to be above 6 percent. Today, it may be closer to 5.5 percent. I will use the value 5.5 percent in my examples.

Why does inflation usually rise if the economy is below the CIRU? A low unemployment rate raises the relative bargaining strength of workers. The fear of layoffs diminishes, and unions become more aggressive. Employers have difficulty finding workers and are afraid of losing workers through stingy wage increases. Moreover, a low unemployment rate usually means that businesses are enjoying high revenues and profits, so they can afford higher wage increases. Thus, a low unemployment rate usually results in a rise in wage increases, cost increases, and price increases—that is, a rise in inflation.

Just the reverse is true if the economy has a high unemployment rate. A high unemployment rate reduces the relative bargaining

strength of workers. The fear of layoffs rises, and unions become more timid. Employers have no difficulty finding workers, and have no fear of losing workers through stingy wage increases. Moreover, a high unemployment rate usually means that businesses are suffering low revenues and profits, so they are forced to reduce wage increases. Thus, a high unemployment rate usually results in a fall in wage increases, cost increases, and price increases—a fall in inflation.

Imagine for a moment that inflation is zero and unemployment is at the CIRU—roughly 5.5 percent. How can the Fed keep it that way? If output per worker—labor productivity—grows 2 percent per year due to capital accumulation and technological change, then the same number of workers can produce 2 percent more output next year. So if real output grows 2 percent, the unemployment rate will stay at 5.5 percent if the labor force—the number who want to work—remains constant. If the labor force grows 0.5 percent instead of staying constant, then real output must grow 2.5 percent to keep the unemployment rate at 5.5 percent. I will call the sum of productivity growth and labor force growth *normal real output growth*. In this example, normal real output growth is therefore 2.5 percent (2 percent plus 0.5 percent).

Suppose the Fed chooses an ambitious goal of 0 percent inflation. Then the Fed's job is to make total dollar spending in the economy rise 2.5 percent per year. If it does, there will be enough demand to buy 2.5 percent more real output, since 0 percent is needed for price increases. If real output rises 2.5 percent, then the unemployment rate will stay at 5.5 percent. And if the unemployment rate is 5.5 percent, then experience shows that inflation should stay steady; if the inflation rate was 0 percent last year, it should stay 0 percent this year.

So if the Fed wants a 0–inflation economy with the unemployment rate at the CIRU, it must try to keep the growth rate of total spending equal to normal real output growth—2.5 percent in this example. However, for the reasons given earlier in this chapter, the Fed may prefer a 2 percent inflation rate. If so, then its task is to keep the growth rate of total spending at 4.5 percent, so that 2 percent of the spending goes for price increases, and 2.5 percent for an increase in real output.

We can summarize: The Fed's target is to achieve a growth rate

of total spending equal to normal real output growth plus the desired inflation rate. Since normal real output growth is roughly 2.5 percent, if the Fed accepts an inflation goal of 2 percent, then the Fed's normal target should be a growth rate of total spending of about 4.5 percent.

The Great Offsetter

How can the Fed meet its target for total spending growth? The answer is that the Fed is the great offsetter. Its job is to assess the other forces influencing total spending in the economy and to lean against the wind.

Suppose the Fed estimates that household and business psychology and government budget policy will all tend to make spending grow too slowly in the months ahead. Then the Fed's job is to stimulate spending. How? By reducing interest rates. How? By injecting more cash into the banking system. How? By buying more government bonds in the open market.

Recall again how this works. When the Fed buys government bonds, the sellers—households, businesses, and governmental units—deposit Fed checks at their banks. The banks, in turn, obtain cash from the Fed to cover the checks. The banks then seek borrowers so that they can earn interest on part of this cash. The more bonds the Fed buys, the more cash the banks obtain, and the lower the interest rate they must offer to induce potential borrowers to take all the cash they seek to lend.

So to stimulate more borrowing and spending in the economy, the Fed simply raises its *open market* purchases of government bonds. It does just the reverse if it wants to reduce the growth of spending in the economy.

Of course, hitting it just right is no simple task. I don't want to give the false impression that the Fed can fine-tune the growth of total spending. After all, it's hard to guess at consumer and business psychology. Currently, it's difficult to guess what government budget policy will be. And even if the Fed knew these precisely, it could not be sure exactly how much to adjust its bond purchases because the link between money injection and the fall in interest rates and rise in spending is imprecise.

So we can't expect perfection from the Fed. But the Fed is often able to do a pretty good job. Its excellent technical staff uses the best econometric models—economic models based on statistical analysis of U.S. economic data—to try to estimate the bond purchases that will keep total spending growth on target. And it continuously adjusts its open market operations based on new data.

The Fed has controlled the economy fairly well since the deep recession of 1982. Remember, it intentionally engineered the 1982 recession in order to bring down inflation. Since then it has kept spending growth near its target, so that usually the unemployment rate has stayed near 5.5 percent, real output growth has stayed near 2.5 percent, and the inflation rate has stayed steady below 3 percent.

However, the Fed has done better in some years than in others. Under the leadership of its new chairman, Alan Greenspan (appointed by President Reagan), it handled the stock market crash of October 1987 very nicely, so that the economic recovery was hardly interrupted. The Fed missed at the beginning of the 1990s, and there was a recession, with the unemployment rate rising from 5.6 percent in 1990 to 6.8 percent in 1991 to a peak of 7.5 percent in 1992. This was much milder than the intentional recession of 1982, with a peak unemployment rate of 9.5 percent that lasted two years. But it was severe enough to help defeat President Bush in his bid for reelection in 1992. In the 1990s, the supply curve of the economy gradually shifted down due to cost-reducing information technology, simultaneously raising GDP and reducing inflation, making the Fed's job easier. The advances in technology were genuine, but many people got carried away with excessive enthusiasm, and "irrational exuberance" (Chairman Greenspan's famous phrase) caused the stock market to soar. But in 2000, many sobered up, the stock market plunged, and the economy dropped into a recession. The Fed moved quickly to combat it by cutting interest rates. But as of mid-2003 the economy still hadn't strongly recovered.

So what can we conclude? When it comes to recession, the Fed is our first line of defense, but it can often use some help from Congress and the president, and can sometimes get it. The $600 tax rebate enacted by Congress in 2001 helped the Fed counter the recession. But when it comes to inflation, the Fed is probably our only line of defense. It would be risky to depend on Congress and the president

to take the unpopular actions needed to reduce demand—reducing government spending and/or raising taxes. To bring down inflation, the Fed is probably both our first and last line of defense.

Reducing the CIRU

Once inflation is at its target, the Fed should try to keep the unemployment rate at the CIRU, which currently appears to be about 5.5 percent in the United States. But can anything be done to reduce the CIRU of the economy? If the CIRU could be cut, for example, to 4.5 percent, then the Fed could safely reduce the unemployment rate to 4.5 percent without generating a rise in inflation. The answer is that the CIRU *can* be reduced—it's not "natural" (which is one reason I like the term CIRU better than "natural rate of unemployment"). Some methods of reducing the CIRU should receive broad public support provided the cost is reasonable: for example, government funding of job placement services so that there is a quicker matching of unemployed workers and available job vacancies; or funding of job training programs for workers displaced through no fault of their own (for example, from international trade). But other methods of reducing the CIRU are controversial: for example, reducing the level of unemployment benefits, which would compel unemployed workers to take any new job quickly.

Thus, the citizenry must weigh the costs as well as benefits of policies that can reduce the CIRU. Note that policies to reduce the CIRU are in the hands of Congress and the president, not the Fed. The Fed can't reduce the CIRU. The Fed's job is to reduce inflation to its target (for example, 2 percent), and then keep the actual unemployment rate of the economy close to whatever CIRU Congress and the president have achieved.

7

Growth

Senator Myopia had the mall crowd cheering. Outside, the snow fell, and a weaker people might have huddled in their homes. But undaunted, sturdy Americans had ventured out into the winter storm, lured by the warmth of the great American mall. Now in the warm belly of the mall, surrounded by sparkling shop windows, how glad they were that their pioneer fortitude had triumphed.

The shop windows of the great American mall vibrated as the crowd roared its approval. Only a speaker like Senator Myopia could grab the attention of frantic mall shoppers and interrupt their frenzied purchases.

"I'm sick and tired of the party poopers, the killjoys, the austerity pushers, and the discipline devotees who say we must sacrifice to stay number one. How dare they attack our most sacred institution, the great American mall!" roared Senator Myopia.

"You know, I've traced my lineage all the way back to ancient Greece and Rome, but my fellow Americans, let me tell you something. The Romans were great builders, but their empire fell, and do you know why? Because they never built a great shopping mall. And this, my friends, is why America shall endure and thrive forever. Oh sure, we must save more. Sure, we must invest more. But first and foremost, my fellow Americans, we must consume more!" The crowd erupted in thunderous applause.

"Let's make those cash registers hum," continued the senator. "The more we consume, the stronger our economy will be. Who dares to spoil our party?" mocked the senator as the crowd broke into approving laughter.

"I do," said a young man in the corner. The crowd hushed, and a thousand eyes turned to him. He began to speak, but Senator Myopia immediately interrupted him.

"So you'd like to spoil our party, would you?" asked Senator Myopia with a confident grin. "And you expect us to listen to you. But just look at you. Your clothes. Who handed them down to you, your older brother or your father?" The crowd began to snicker.

"And there's nothing in your arms, except a single book. Where are your packages? Could it be you left your credit cards home? Or don't you even have a set of credit cards?" The very thought of someone without credit cards sent the crowd into a fit of laughter.

But the young man seemed unruffled. In a calm voice he said: "You would be right, Senator Myopia, if our economy were in a recession. Then we would need households to spend more on consumer goods, and businesses to spend more on investment goods. In a recession, it would be possible to produce more consumer goods and investment goods at the same time, because workers are unemployed and factories are idle. Some unemployed workers can go to work in factories that make consumer goods, while other unemployed workers can go to work in factories that make investment goods."

"You see," said Senator Myopia triumphantly to the crowd, "we can consume more goods and invest more in machinery at the same time!"

"But I'm afraid," continued the young man, "that once our economy has recovered from recession and virtually everyone who wants to is working, it is only possible to make more investment goods by making less consumer goods. Some of the workers making consumer goods would have to be shifted to factories making investment goods."

"What do you mean?" asked Senator Myopia angrily.

"Once we have made our economic pie as big as possible by employing everybody, then the only way to get more investment goods is to take a smaller slice of the fixed pie for consumer goods. To raise the investment slice from 20 percent to 24 percent, we must reduce the consumption slice from 80 percent to 76 percent."

"What are you talking about?" exclaimed Senator Myopia.

"Another way to say this," said the young man calmly, "is that we must be willing to raise our saving rate from 20 percent to 24 percent to make our investment rate rise from 20 percent to 24 percent. As Eve explained to Adam in chapter 1 of this book that I'm holding in my hand, investment equals saving, so to raise investment, we must raise saving, and that involves sacrificing some consumption."

Now Senator Myopia became angry.

"Who are you?" he shouted at the young man. "What do you do for a living?" muttered voices in the crowd.

"I'm an economist," answered the young man in a quiet voice.

"An economist!" Senator Myopia smirked. "Did you hear that, my fellow Americans? An economist! Is this the kind of person you would invite to your house for dinner? Would you want your children around this kind of person? What do you do for fun, young man, read a textbook?" The crowd went wild with derisive laughter.

"Ignore him, Senator!" shouted the crowd. And so the senator did. He brought his oration to a dramatic crescendo.

"Consume, consume, consume, my fellow Americans. Let's go out and spend ourselves rich!" With that, the crowd cheered and burst into the waiting shops. The senator and his party soon left, triumphant.

But in the corner, the young man remained. And he did not remain alone. Several people, young and old, quietly gathered around him, and urged him to teach them.

"Before I can begin to teach you," said the young man, "you must first read the parable of Adam and Eve and the tractor."

"But all of us have read the parable," said the listeners in unison.

"I know you have. But that was a few weeks ago. To follow what I'm about to teach you, you must refresh your memories. Do you remember exactly how Eve showed Adam that investment must equal saving?"

"Not exactly," said the listeners in unison.

"I've brought you each a copy of the book with the Adam and Eve parable. Let's take ten minutes right now to reread it."

And so they did.

When they finished rereading the parable, the young man began to teach, and here is what he said.

Capital Accumulation

I'm afraid, my friends, that I've begun with a simple, unpleasant truth. A higher investment rate requires a higher saving rate, because investment equals saving, and a higher saving rate inescapably means a lower consumption rate, because they must sum to 100 percent. So, should we raise the saving rate? Yes, I believe there is a decisive reason for raising our saving rate: the relative standard of living of our children and grandchildren. If we do not raise our saving rate, then within a few decades, several other nations will overtake our standard of living. If we maintain our current saving rate, today we will enjoy the highest consumption per person in the world, but tomorrow our children and grandchildren will not.

Senator Myopia mocked my clothes. He accused me of not caring about consumption. But that is untrue. I do not come to preach against materialism. Far from it. I love CDs, DVDs, airplane travel, and many other material things. I do not place myself above my fellow citizens who shop frantically all around us. My difference with Senator Myopia is simply this: In his obsession with consumption today, he forgets about consumption tomorrow. He fails to grasp what we must do today to protect consumption tomorrow. He thinks the best way to achieve high consumption tomorrow is to enjoy fast consumption growth today. Unfortunately, he is wrong. So I come to plead for a higher saving rate today, not because I am against consumption, but because I am for it—in the future as well as the present. I want our future consumption to be second to none in the world.

Nor am I an extremist. I don't advocate an actual cut in our consumption, only slower consumption growth for a few years. Instead of our normal 2.5 percent consumption growth, let it grow 1.5 percent for half a decade. I want a small cut in our consumption *rate* (the fraction of our output that we consume) each year for several years— small enough so that our dollar consumption keeps rising each year, yet more slowly than it would otherwise.

To understand my case for gradually raising our national saving rate, you must first understand what capital accumulation is, and how it raises the standard of living (the level of consumption per person).

Let me begin with a question. What determines the rate of improvement in a nation's standard of living in the long run? Of course,

many factors influence the rate of advance. But one source, my friends, deserves the spotlight: capital accumulation. What is capital accumulation? Are you thinking of stocks and bonds, those impressive pieces of paper we often lock up in a bank vault for safety? Or are you thinking of financial capital, the funds that finance the purchase of stocks and bonds? By "capital," I do not mean either the pieces of paper or the funds that buy them. Instead, I mean *real* capital, which, when combined with labor, produces real output. Capital enables the average worker to produce more output.

What do you think real capital is? Are you thinking of machines? Many people think capital consists solely of physical capital, like machines. But this view of capital is too narrow. Of course, physical capital is vital. What would our standard of living be without machinery, factories, roads, and bridges? Where would today's farmer be without a tractor?

But capital is more than machinery. Capital is also the stock of technical knowledge accumulated from past experience. This stock of "blueprints" tells us how to produce specific goods and services. Just imagine the consequence of a national amnesia that would require us to reinvent the wheel and everything else. Did you realize that capital is also the skill of the labor force that is acquired by education and training? The stock of blueprints and machinery are not very effective unless the workforce has accumulated the human capital (skills) needed to follow the blueprints and operate the machines.

Suppose the capital stock is $2,000 billion on January 1 and that during the year $500 billion of new capital goods are produced and $300 billion of old capital goods wear out—*depreciate*—so that the capital stock is $2,200 billion on December 31. Then we say that this year's *gross investment is* $500 billion and *net investment is* $200 billion. So *net* investment is the net increase in the capital stock that occurs during the year—in this example, $200 billion. In a given year, capital accumulation equals net investment.

Now imagine a simple economy that produces only two goods: corn, the consumption (C) good; and tractors, the investment (I) good. Assume that all available labor, capital, and land will be utilized to produce either corn or tractors. The economic year is beginning. How much labor, capital, and land will be assigned to make tractors, and

how much to make corn? Clearly, more labor, capital, and land for tractors means less labor, capital, and land for corn. The tractor production sector—the I sector—can only expand if the corn production sector—the C sector—contracts.

How nice to imagine a world with only corn and tractors. But let's return to the real economy. What belongs in the C sector? The I sector? Obviously, a wheat farm is in the C sector, and a tractor factory is in the I sector. But what about a school? A school should also be in the I sector because it produces human capital. So should the research and development division of every business firm, because the division produces knowledge capital. Capital accumulation increases when a greater share of production occurs in the I sector, and less in the C sector.

Each unit of input—land, labor, or capital—can be assigned to produce goods and services that will be used up—consumed—this year, or to produce goods and services that raise the productive power of workers in the future. The more units of input that are assigned to produce C goods, the fewer that are available to produce I goods.

Imagine a circular pie, representing national output, that is divided into two unequal parts. The large slice is consumption, and the small slice is investment. The only way to increase the investment slice is to reduce the consumption slice. A higher investment rate (percentage) requires a lower consumption rate (percentage). If more land, labor, and capital are devoted to the production of investment goods, less must be allocated to the production of consumer goods.

So I'm afraid we can't escape a painful truth, no matter what Senator Myopia says. Capital accumulation requires a sacrifice in the present. In order to build machinery (physical capital), improve our skills (human capital), or invent new technology (knowledge capital), time and resources must be diverted away from producing goods and services for current consumption. But does capital accumulation really raise future *productivity* (output per worker)? Let's use some common sense. Why is the productivity of the average American worker today so much higher than the productivity of the average American worker one hundred years ago? Is it because our great-grandparents were lazy, and we work hard? Nonsense.

The central reason is that today's American worker has more education and skill (human capital), utilizes more and better machinery

and technology (physical capital), and follows a more advanced set of blueprints (knowledge capital).

Compare the primitive farmer, who lacks both a tractor and the skill to operate it, with the modern farmer, who possesses both the tractor (physical capital) and the ability to operate it (human capital). Moreover, the modern farmer utilizes knowledge capital accumulated from past experience, research, and invention to tell him which farming techniques will be most productive. Is it any wonder that output per worker—productivity—is much higher for the modern farmer?

Countless urban and industrial as well as rural examples make the commonsense point: Raising capital per worker generally raises output per worker. Raising capital per worker is therefore the key to raising the standard of living, or consumption per person.

So far I have said nothing about saving. I have explained why capital accumulation through investment is the key to advancing the standard of living, and why raising the investment rate requires reducing the consumption rate. But what about saving? More investment requires more saving, for the simple reason that investment must equal saving. Why?

Saving is defined as income not consumed. Investment is defined as output not consumed. But income must equal output, because for every dollar of output sold, a dollar of income is earned. If output and income are $1,000 billion, and consumption is $900 billion, then income not consumed—saving—is $100 billion; and output not consumed—investment—is also $100 billion.

Keep this simple example in mind. To finance the $100 billion of investment—the purchase of machinery—imagine that business firms issue $100 billion of bonds, and savers purchase the $100 billion of bonds. In effect, business firms borrow $100 billion from savers (lenders) to invest in $100 billion worth of machinery. The machinery generates a real return—it raises output. Firms use the additional revenue earned on the machinery to pay interest to savers (bondholders). In this example, we can say that the $100 billion of saving is necessary to *finance* the $100 billion of investment. National income is $1,000 billion. Households choose to consume 90 percent ($900 billion) and save 10 percent ($100 billion). This saving is what makes possible the $100 billion of investment.

So the only way to raise capital accumulation is to raise saving. Raising the national saving rate—the fraction of national income that is saved rather than consumed—is the key to raising the future standard of living.

The Saving Rate Must Be Raised Gradually

But here we are, surrounded by shop windows displaying consumer goods, and I am claiming that raising our saving rate will raise our future standard of living. How can I claim that? If we raise our saving rate, we reduce our consumption rate. But won't this mean less spending at the mall? And won't the mall cut its orders from manufacturers of consumer goods? And won't these manufacturers cut production and lay off workers? Economists call a fall in output and employment a *recession*. So won't the result be a recession?

Yes, there would be a recession if we raised the saving rate suddenly and sharply. But there need be no recession if we raise it gradually. Why?

Here's the key point. Today, output, consumption, and investment all grow at roughly 2.5 percent per year. Our aim is to make consumption grow more slowly—say 1.5 percent per year instead of 2.5 percent—while we make investment grow more rapidly, for about half a decade. If we gradually raise the saving rate, dollar consumption will simply grow more slowly, but it will never actually decline. Thereafter, consumption, investment, and output will all grow somewhat faster than 2.5 percent due to the permanently higher investment rate.

When I say we must reduce the consumption *rate* (the fraction of output that we consume) to raise the investment rate, this doesn't mean that dollar consumption must literally decline from one year to the next. So the mall will never suffer a decline in sales, only a slower growth in sales.

What will happen if the saving rate is raised gradually, or equivalently, if the consumption rate is reduced gradually? As workers voluntarily quit and retire in the consumption (C) goods sector, they will not be replaced. Most new jobs will open up in the investment (I) goods sector. So layoffs will be avoided in the C sector.

Of course, if we were foolish and tried to shift the relative size of

the consumption and investment goods sectors too quickly, then we would need to force workers out of the C sector. The result would be layoffs. But by raising the saving rate gradually, we can avoid layoffs. Voluntary job-leaving (quits) and retirements will handle the required contraction in the C sector's workforce.

But can we be sure that jobs in the I sector will expand enough to prevent a rise in unemployment? As the great economist John Maynard Keynes emphasized in his classic, *The General Theory of Employment, Interest, and Money* (1936), when household demand for consumer goods grows more slowly, this does not guarantee that business demand for investment goods will grow more rapidly. But unless it does, the economy will not generate enough new jobs, and unemployment will rise.

Here is where our central bank, the Federal Reserve ("the Fed") comes in. It is the Fed's job to make sure that the investment goods sector grows more rapidly when the consumer goods sector grows more slowly, so that enough new jobs are created in the economy to prevent unemployment from rising. How can the Fed do this?

It's simple. The Fed can reduce interest rates. When interest rates fall, business firms throughout the economy are encouraged to borrow to buy machinery and new technology. They raise their demand for investment goods. The lower the interest rates, the greater the demand for investment goods. In turn, the firms making I goods will need more workers to meet the increased demand. So more new jobs will be created in the I sector.

But how does the Fed lower interest rates? By injecting more money into the economy and the banking system. How? By *open market operations*, or buying government bonds from households, businesses, or local governments. The sellers of bonds deposit the money in banks. In response to the infusion of cash reserves, banks try to increase lending. To attract borrowers for their excess funds, banks compete by reducing interest rates. At lower interest rates, business firms find it profitable to borrow more to buy more investment goods. So the Fed's action results in an increase in investment demand by business firms.

John Maynard Keynes wrote a brilliant book in the 1930s, but it has been misinterpreted. Many people have come to believe that saving hurts the economy. It is simply not true that an increase in saving

must cause a recession. It is only true if the Fed fails to implement routine, appropriate, offsetting monetary policy. Once again, what must the Fed do? When the demand for consumer goods grows more slowly, few new jobs are created in the C sector. Hence, the Fed must make sure that demand for investment goods grows more rapidly so that enough new jobs are created in the I sector. The Fed can easily do this by reducing interest rates, thereby encouraging business firms to borrow to buy more I-sector goods.

But this is all theory. Does it work in practice? Look at other countries. For four decades, from 1950 to 1990, Japan had a much higher saving rate than the United States but seldom had a recession (true, Japan went into recession in the 1990s, but for other reasons—a plunge in its overpriced stock market and real estate market, and problems in its banking system). Several European countries also have had higher saving rates. Yet these high-saving countries have not had more recessions, or higher unemployment rates, over these decades. So the theory works in practice. A higher saving rate does not imply a higher unemployment rate.

Let's sum up. If the saving rate is raised gradually, and the Fed earns its pay by implementing proper monetary policy, then our economy can adjust to a higher saving rate without a recession or a rise in unemployment. The higher saving rate means that a larger fraction of our output will be I goods instead of C goods.

So you thought you were helping the economy by consuming? You said, "Who will buy goods, if not us? And if we don't buy goods, producers won't make them. And if they don't make them, workers will be laid off. And there will be hard times. So we are patriots when we consume, and traitors when we save."

But now you see your error. You are guilty of the sin of pride. You consumers are not the only buyers of goods in the economy. Business firms buy goods—investment goods. The correct way to look at it is this: The more consumer goods you demand, the more consumer goods producers will make, and the fewer resources—labor, capital, materials, and land—will be available to make investment goods. So you are directly competing with business firms. More consumer goods for you means fewer investment goods for them.

You thought you were heroes, the only buyers in town. Without you, you thought, nobody would buy goods, production would plum-

met, and down would go the economy. What pride! But you are not the only buyers in town. Business firms are also buyers. And your buying of consumer goods interferes with their buying of investment goods, because the total output of goods that can be produced in a given year is limited.

So don't flatter yourselves. The economy's health does not depend on fast growth in your consumer spending. If you slow the growth of your consumer spending, the Fed will make sure that business firms quicken the growth of their investment spending, and total spending will still grow normally. But now more of the growth in output will be in investment goods, and less in consumer goods.

So, if some economist sets before you a policy that will encourage saving and discourage consumption, do not tremble for the fate of the economy. Such a policy is exactly what is needed to raise the future productive power of the economy. As long as the policy is phased in gradually, so that it raises the saving rate gradually, do not fear it, but welcome it. Such a policy is economic medicine that is safe and effective.

Fiscal Discipline and Growth

We've seen that economic growth takes discipline: people must be willing to consume less and save more in the short run, in order to generate more productive power in the long run. But people aren't the only ones who must show discipline. So must the government. Another term for government discipline is *fiscal discipline*. Fiscal discipline means that the government limits its borrowing, except in a recession (where, as explained in chapter 5, the government *should* run a deficit and borrow). But why does fiscal discipline promote economic growth?

The reason is simple: The more funds the government borrows from the public, the less funds are available for businesses to finance investment in real capital. Suppose people save $1,000 billion. If the government doesn't borrow any of it, people can deposit the $1,000 billion in their banks, and the banks in turn can lend the $1,000 billion to business firms to purchase new capital equipment; so the $1,000 billion of saving finances $1,000 billion of investment. But if the government borrows $400 billion from savers (be selling them $400 billion of government bonds), then savers will only deposit $600 bil-

lion in their banks, and the banks will lend only $600 billion to business firms to finance investment. We can say that government borrowing "crowds out" investment.

To avoid borrowing, the government must limit its spending to what it raises in tax revenue—it must balance its budget. Suppose the government budget is balanced, and a senator says, "Let's raise government spending." Then to protect investment and economic growth, we should insist that the spending increase be financed by a tax increase, not borrowing. Similarly, suppose another senator says, "Let's cut taxes." To protect growth, we should insist that the tax cut be matched by an equal spending cut, so the government doesn't have to borrow.

But I can hear Senator Myopia. He went around the country blustering, "Who says we can't raise our consumption and investment at the same time? I know how to do it. Let's cut taxes, but don't worry, I won't cut government spending—I don't want to cut your favorite programs. My tax cut will still work, though. If I cut your taxes $100, you'll consume $90 and save $10. More saving means more investment, so we'll get more of both—consumption and investment!"

Was he right? Unfortunately, he was not. But where is the mistake? Let's see if we can find it.

First of all, he was right about one thing: If your taxes are cut, you'll consume more, and if you consume more, then the C-goods sector will raise its production. But he ignored the fact that the C-goods sector will then have to draw labor, capital, and land away from the I-goods sector, so production of I goods will be forced down. More corn means fewer tractors. If taxes are cut $100 billion, and households consume $90 billion more, then the I sector will be forced to produce $90 billion less.

True, if taxes are cut $100 billion, households might raise their saving $10 billion. Doesn't more saving mean more investment? So where is the mistake?

Yes, *national* saving must equal national investment. But national saving is the sum of household, business, and government saving. The tax cut will raise household saving. In fact, in our example, households save 10 percent of the tax cut, or $10 billion. The mistake is forgetting about the impact of the tax cut on another component of national saving—government saving.

What is government saving? Saving is always income minus consumption. Therefore, government saving is government net income minus government consumption. Government net income equals tax revenue minus cash "transfer" payments to households or business firms, such as Social Security benefit payments. Government consumption is the purchase of goods or services by government that yields current rather than future benefits to citizens—for example, the purchase of the labor services of recreation workers on public playgrounds. Actually, most government purchases are investment rather than consumption because the goods and services yield future rather than current benefits. For example, the purchase of labor services for the construction of highways or schools, or the purchase of military goods such as tanks, yields benefits primarily in the future. So government saving is determined mainly by government net income—taxes minus transfers.

Now that we have the spotlight on government saving, we can illuminate Senator Myopia's error. Senator Myopia is holding government spending constant; he wants you to have all your favorite programs, including transfers and government consumption. So what happens to government saving when taxes are cut $100 billion while government spending is held constant? While households cheer the $100 billion tax cut, the poor government treasurer is despondent: government net income falls $100 billion, and since government consumption stays constant, government saving falls $100 billion.

Since household saving increases $10 billion but government saving falls $100 billion, national saving falls $90 billion. So national investment must also fall $90 billion. And this is exactly the same answer we obtained before by recognizing that consumption would increase $90 billion, forcing investment down $90 billion.

Then what is Senator Myopia's mistake? He forgets that household saving is not the same thing as national saving. A $100 billion tax cut may raise household saving $10 billion, but it reduces national saving $90 billion, and therefore, reduces national investment $90 billion.

Am I claiming that any tax cut must reduce national saving? Not at all. Senator Myopia overlooked one little detail. He forgot to match his tax cut with an equal cut in government transfer spending. What

would have happened if the $100 billion tax cut were matched by a $100 billion cut in government transfer spending?

The answer is simple. There would have been no change in national saving. Taxpayers would have consumed $90 billion more, but transfer recipients would have consumed $90 billion less, so there would have been no change in total consumption. Taxpayers would have saved $10 billion more, but transfer recipients would have saved $10 billion less. Government net income—taxes minus transfers—would have been unchanged, so there would have been no change in national saving.

Thus, an equal cut in taxes and government transfers has no effect on national saving and investment. What hurts national saving and investment, however, is a tax cut that is not matched by an equal cut in government transfer spending. Shame on you, Senator Myopia.

Of course, the same thing would happen if the government raises transfers $100 billion but doesn't raise taxes. Consumption would increase $90 billion, so there would be $90 billion less resources for investment. Private saving would increase $10 billion, but government saving would fall $100 billion, so national saving would fall $90 billion, and so would national investment.

Thus, national saving and national investment fall if taxes are cut without cutting government transfers, or if transfers are increased without raising taxes. Hence, our message to conservatives is, "Don't cut taxes unless you are willing to cut transfers by the same amount," and our message to liberals is, "Don't raise transfers unless you are willing to raise taxes by the same amount."

Old Karl's Mistake

We can learn about the importance of capital accumulation from history. Why did the average American in 1900 have a much higher standard of living than the average American in 1800? Why is the average American of 2000 much better off than the average American of 1900? Why was poverty the rule two hundred years ago, but the exception today in the United States? For that matter, why is it the rule today in some countries, but the exception in others?

A dramatic story in the ascent of man began some two hundred years ago: the Industrial Revolution. Today, with hindsight, it is clear

that it eventually resulted in a drastic reduction in poverty. But we would do well to remind ourselves that the early reactions were, to put it mildly, mixed. The Luddites did not exactly welcome machinery. And with good reason. Here were profit-driven entrepreneurs replacing people with machines. Machines were not poverty reducers, said the Luddites; they were poverty creators. Where would it lead? Greedy capitalists would keep substituting capital for labor. Unemployment would rise, and workers would be impoverished. Incidentally, the same argument was made about automation in the 1960s. And many believe it today.

There is one little problem with this logic. How is it that today, with so much more machinery per person, the unemployment rate is no higher than two hundred years ago, and the average worker is so much better off? There must be a flaw in the logic. What is it?

Here's the mistake. The Luddite argument really says this: "If you're going to produce the same output but now use more machines, then you're going to need less labor." Correct. But who says we're going to produce the same output? Suppose instead that we add machines, but keep the same labor. Then we'll get more output. And that is, in fact, what happened over the long run for the economy as a whole.

Of course, it didn't always happen in the short run at a particular factory, which is how the argument got started that machinery impoverishes workers. In the short run, the workers at a particular plant were not dreaming. Some really were laid off because of the introduction of machinery. Why? Why couldn't the employer simply keep all the workers and raise output? Because in the short run, the market for this particular product may not have supported so large an increase in its output. So in some cases, the introduction of machinery did cause layoffs. And workers were unemployed until they could find jobs elsewhere.

To understand why it all worked out well in the end, let's consider the argument of someone who is still remembered today for predicting otherwise: good old Karl Marx. Many people think that old Karl disputed every claim made by advocates of capitalism. But this is untrue. He only took issue with the last claim: that eventually capitalism would pull workers out of poverty.

Marx fully agreed that machinery raises the productivity of

workers—output per person. He conceded that profits were often saved, that saving financed investment in machinery, that machinery increased the output of each worker, and that this could *potentially* enable the capitalist to raise the wage, and hence consumption, of the average worker. But Marx denied that the wage would in fact be raised enough to enable workers to escape poverty.

Instead, he contended that the greedy capitalists would hold down the wage despite the worker's higher productivity. To be fair to old Karl, he may have conceded that some rise in the absolute wage might occur. But he insisted that the wage of workers would deteriorate relatively if not absolutely. The failure of the wage to keep pace with worker productivity would, Marx predicted, lead to less than complete happiness among the masses. You know the rest. Now at certain times and places, old Karl was right. The rise in worker productivity, made possible by machinery, was not matched by a comparable rise in the wage. But over the long haul, it is now clear that it would be hard to make a more inaccurate prediction. In fact, the real wage and consumption of today's worker in capitalist economies is dramatically higher than it was 150 years ago. The rise in worker productivity has eventually led to a significant rise in the worker's wage and standard of living.

So how did this happen? Why did the worker's wage eventually rise with his productivity? Most economists, myself included, give the answer illustrated in Figure 7.1. Let's begin with an overview of the story. The demand for labor comes from employers, and the supply of labor comes from workers. Initially, the D and S curves intersect at a wage of $1.00. As employers accumulate more machinery, they demand more workers to operate the machinery so the D curve shifts right. At $1.00, there is excess demand for labor, so each employer competes for labor by raising the wage. Consequently, the wage is bid up to $2.00. Hence, capital accumulation by employers ends up raising the wage of labor.

Now let's tell the Figure 7.1 story in more detail. Imagine you're a profit-seeking employer back in the good old days. How do you decide how many workers to hire? Figure out the additional output, and hence additional revenue, that another worker will give you. Compare it to the wage you must pay him. If the additional revenue exceeds the additional cost, hire him, and make the same comparison

Figure 7.1 **Rise in the Wage**

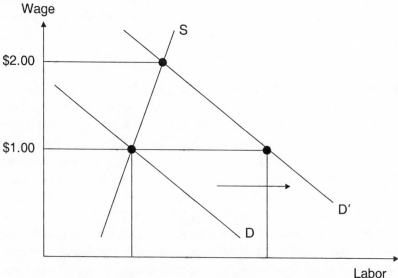

for the next worker. Of course, as diminishing returns sets in, it eventually does not pay to hire any more at the going wage. Suppose, like a good profit-seeking capitalist, you decide to stop at the hundredth worker, because he adds only slightly more than $1 of revenue per hour, and the going hourly wage is $1.

Now you save some of your profit and use it to invest in new machinery, which raises the productivity of your workers. Suppose that the revenue you obtain from your hundredth worker is now $2, instead of $1. In fact, you calculate that, with your new machinery, you could hire another twenty workers (for a total of 120) before diminishing returns makes another worker's revenue fall to $1.

What do you do? If the going wage is still $1, you try to hire twenty more workers. But now comes a key part of the argument. Assume that there are many other capitalists, and they are doing the same thing you are, so they all want to hire more workers, but most workers are already employed. With only so many workers available, you can't all get what you want. At a $1 wage, the total demand for labor in the economy now exceeds the total supply. So what do you do?

Naturally, you try to bid workers away from other capitalists. You

do it by offering a wage slightly above $1, say $1.10. When other workers hear of your offer, some are glad to switch. Of course, other capitalists will not sit idly by and watch their workforce, and profit, disappear. Because they too have introduced machinery, they too will find it profitable to match you, and even raise you. They may raise their wage offer to $1.20, not only winning back their workers, but luring a few of yours. So now what do you do?

Remember, if necessary you can go all the way to $2 to try to keep your original hundred workers because your machinery has doubled the productivity of your hundredth worker. But you really don't want to do this. You would like to get all capitalists together in a room and make a speech: "Let's not engage in a foolish wage competition. Where will it get us? Let's keep the wage at $1, be satisfied with the same workforce, and enjoy a nice profit."

But you are a practical capitalist, and you realize that your speech won't work. Why? Because whoever cheats and offers a slightly higher wage will reap even more profit. Every capitalist will realize this, and everyone will be tempted to cheat. If you are foolish enough to hold the line on wages, you will simply lose your workforce and all your profit. Besides, suppose some capitalist is caught cheating. What can you do to him? Expel him from the local capitalist society? He will laugh all the way to the bank.

So you've got no choice. You've got to engage in a wage competition to hold on to your workforce. So sure enough, the wage goes to $1.10, $1.20, and eventually nears $2. There it stops, because at that wage, finally, the typical capitalist does not want to expand his workforce. For example, in your case, your hundredth worker adds revenue of just $2, so it wouldn't pay you to hire more than a hundred if the wage hits $2. So when the wage reaches $2, total demand for labor again matches its supply, and the wage stops rising.

You and the other capitalists shake your heads and mutter, "What a shame. Oh, if only we could have kept the wage at $1, just like good old Karl said we would." You and other capitalists will probably curl up at the fireplace with a copy of good old Karl's *Das Kapital* and try to console yourselves with the story of what was supposed to happen to the wage.

Now notice a key feature of the economist's explanation. It does not, in any way, assume that the wage rises because capitalists aban-

don, or even moderate, profit seeking. Just the opposite. The argument assumes that capitalists are motivated solely by profit seeking. But it is exactly this profit seeking that makes an agreement against wage competition impractical. Sure, it would be better for all capitalists if they all stuck to the agreement. But whoever breaks it first will make even more profit in the short run. And so, with many capitalists, the agreement simply won't hold.

Of course, few economists insist that this explanation is the whole story. Sure, there were probably some capitalists who thought their workers deserved a wage increase. And there were no doubt many who liked everything about *Das Kapital* except the ending, and who therefore concluded that it might be prudent for all capitalists to raise wages. And of course, trade unions certainly played a role in raising wages at particular workplaces.

But the economist's central point is this: As long as each capitalist cares enough about his own profit to engage in wage competition, the wage will rise with productivity. So whatever raises productivity will raise the worker's wage, and reduce poverty. And it is capital accumulation that raises productivity. So capital accumulation has long been a vital engine for reducing poverty. It is time to recognize this fact in our public discussion.

The Case for Raising Our Saving Rate

A strong case can be made that our current saving rate is too low. We are a low-saving nation compared to many others. But someone might reply: "Maybe these other countries are saving too much and we're saving just the right amount." So the case does not rest just on comparison with other countries. Instead, the case has two parts: a traditional economic argument, and a novel economic argument.

We begin with the traditional argument. The current U.S. saving rate is not a free market reflection of our preferences between present and future consumption because of several government interventions that reduce our saving rate. What are these interventions?

We indirectly tax future consumption by taxing capital income (interest, dividends, and capital gains) under the current income tax. Capital income taxes reduce the future consumption that can be obtained from a given amount of saving. The choice between future

and present consumption is therefore biased against the future by our income tax.

Another source of distortion is social insurance: Social Security, unemployment insurance, and Medicare. Social Security has made a great contribution to the well-being of the elderly for many decades. But many people surely save less, knowing that Social Security will help out when they retire. Similarly, without unemployment insurance and Medicare, many citizens would save more to prepare for the possibility of being laid off or needing medical care in old age. Thus, government social insurance, while generating great benefits for the citizenry, has almost surely reduced the national saving rate.

A final source of distortion is government saving, one important component of national saving. Government saving is not market-generated, but determined politically by Congress and the president. Politicians may believe it is easier for voters to appreciate a tax cut or benefit increase than to grasp the future gain that will result from government saving.

Now let's turn to the novel economic argument. Any economics text teaches that certain goods are "nonexclusionary": even if someone refuses to pay for the good, we cannot prevent her from benefiting from it. We can exclude you from enjoying a TV if you refuse to pay for it, so a TV is a private good. But if we improve the police protection in your area, we cannot exclude you from benefiting even if you refuse to pay, so police protection is a public good.

Economists agree that the free market will generate too little of a public good because of the freerider problem. Each selfish citizen asks: "Why should I voluntarily pay for the good? If others finance it, they can't keep me from benefiting. But if they refuse to finance it, my contribution will be insignificant." Though not all citizens are selfish, a public good will generally be undersupplied by "the market."

It can be argued that there are at least three public goods that are relevant to the optimality of the U.S. saving rate: (1) the international ranking of the future U.S. standard of living; (2) poverty reduction for low-skilled people willing to work; and (3) our contribution to the "ascent of man" through technological progress. Let me explain what each is, why each is a public good, and how this fact affects the optimality of our saving rate.

Today, the U.S. standard of living still ranks first internationally,

but within a few decades our relatively low saving rate may move us down the ranking. If the typical American wants her grandchildren to live in a nation whose standard of living ranks first, with the psychological, political, and military corollaries of that fact, what can she do? She can of course save privately to provide for her own heirs. But she cannot influence the nation's future standard of living ranking through her own saving. That future ranking is a public good for all Americans because any selfish citizen can reason, "If others save and I don't, our future ranking will remain first, and I will enjoy that fact as much as any saver. On the other hand, if I save and others don't, my sacrifice hardly affects the future ranking."

Many citizens and politicians appear to have considerable interest in whether other countries will overtake us economically and whether our grandchildren will enjoy the most advanced economy on the globe. This implies that the international ranking of the future U.S. standard of living is a public good that many citizens value. Yet "the market" will undersupply it. It is therefore possible that many citizens would judge that they were better off if they were all induced, perhaps even compelled, to save more.

Let us now consider poverty reduction. As we explained above, most economists agree that raising the saving rate will make capital per worker rise faster, and this will make the productivity and the real wage of low-skilled workers grow faster. Hence, raising the saving rate will reduce absolute poverty faster for low-skilled people willing to work. Many citizens seem to value faster poverty reduction for such people. Yet such poverty reduction is a public good. Any selfish citizen can reason, "If others save more so that poverty declines faster, I will enjoy witnessing the reduction almost as much as any saver." So each waits for others to save for this noble purpose.

Finally, we turn to the "ascent of man." Many citizens appear to feel a pride in the quest of mankind to improve its lot and surmount new challenges. Technological progress has been a key ingredient in this historical drama. For millennia, humans have devised new products and new processes. Inventions, innovations, and breakthroughs have lifted mankind in each era. Obviously, there are dangers as well as great benefits, dislocations as well as advances. But most citizens appear willing to keep technological progress driving

forward while at the same time trying to safeguard against potential hazards.

It is likely that raising the saving rate will speed the rate of technical advance and, one hopes, the ascent of man. Surely many citizens
want their own generation to contribute to the ascent through further
technological progress. Yet such a contribution is a public good. A
selfish citizen can think, "If others save more, then technical progress
will be faster and I will enjoy watching mankind's progress almost
as much as any saver."

Let me summarize. A traditional economic argument for raising
the saving rate contends that the current saving rate is less than the
market would generate if there were no government interventions
like capital income taxes, government social insurance programs,
and political incentives for government dissaving. A novel economic
argument for raising the saving rate is that there are at least three
public goods—the international ranking of the future U.S. standard
of living, poverty reduction, and the ascent of man—that are
undersupplied by the market, and that more saving is required to
supply the optimal quantities of these public goods. Together, both
economic arguments provide and build a strong case for policies aimed
at raising the U.S. saving rate.

APPENDIX

The Economy's Long-Run Response to an Increase in the Saving Rate

Suppose we permanently raise our national saving rate. For example,
suppose our saving rate is initially 20 percent (roughly the actual gross
national saving rate of the U.S. economy), and we raise it to 24 percent
permanently. How does the economy respond over the long run?

To answer this question, economists construct a growth model. A
growth model can be very complicated or relatively simple. Let's
consider the answer when a relatively simple model is used. The
model makes several simplifying assumptions. Initially, it assumes a
fixed labor force and no technological change. Capital depreciates,
so new investment is required to maintain the capital stock. In the
initial steady state, the 20 percent saving rate (80 percent consump-

Table 7.1

Response of the Economy to an Increase in Saving Rate

Year	Output/worker ($)	Consumption rate (%)	Consumption/worker ($)
0	1,000	80	800
1	1,000	79	790
2	1,010	78	788
3	1,020	77	785
4	1,030	76	783
5	1,040	76	790
6	1,050	76	798
7	1,060	76	806
Long run	1,110	76	844

tion rate) provides just enough new capital to replace the capital that depreciates. To keep the numbers simple, we assume that in year 0 output per worker is $1,000 and consumption per worker is $800, as shown in Table 7.1.

Now, what happens if the saving rate is raised gradually over a half decade from 20 percent to 24 percent and then fixed permanently at 24 percent? The answer is shown in Table 7.1. In year 1, output per worker remains $1,000; the saving rate is raised to 21 percent, so the consumption rate falls to 79 percent, and consumption per worker falls to $790.

Since a 20 percent saving (investment) rate would have caused new capital to equal depreciation, a 21 percent investment rate causes a small rise in capital per worker and hence in output per worker in year 2. Suppose in year 2 output per worker is $1,010; the saving rate is raised to 22 percent, so the consumption rate falls to 78 percent, and consumption per worker is $788. The 22 percent investment rate again causes new capital to exceed depreciation, and results in another small rise in capital per worker and hence output per worker in year 3. As shown in Table 7.1, beginning in year 4 the consumption rate is fixed at 76 percent, and in year 7 consumption per worker is once again above its year 0 value ($806 vs. $800). Eventually, even with the fixed 24 percent investment rate, diminishing returns cause new investment to no longer exceed depreciation. From then on, capital per worker stays fixed (at a level higher than its initial value),

output per worker stays fixed at $1,110 (11 percent above its initial value of $1,000), and consumption per worker stays fixed at $844, which is nearly 6 percent above its year 0 value of $800. Thus, in this example, society sacrifices for six years (consumption per worker is below its year 0 value of $800) and then is better off beginning in year 7. Eventually, each year consumption per worker is nearly 6 percent higher with the 24 percent saving rate than it would have been with the 20 percent saving rate.

Are we sure that in our actual economy consumption per worker will eventually exceed its initial value? Economists can show that as long as the saving rate is less than the capital (property) share of national income (roughly 30 percent in the United States; labor's share is roughly 70 percent), then consumption per worker will surpass its initial level. In fact, a saving rate equal to capital's share of national income achieves the highest possible consumption per worker in the long run. Thus, it appears safe to conclude that for increases in the saving rate that our economy might actually undertake, it is virtually certain that in the long run consumption per worker will end up higher than its initial value.

Table 7.1 assumes no labor force growth or technological progress, so if the saving rate had remained fixed at 20 percent, output would have remained fixed; consequently, when the saving rate is raised, consumption literally declines through year 4 before rising. But in a more realistic model with labor force growth and technological change, the level of consumption does not literally decline (contrary to Table 7.1) as we gradually raise our saving rate from 20 percent to 24 percent. Suppose labor force growth is 0.5 percent per year and technological change is 2.0 percent per year. Then output, consumption, and investment all normally grow about 2.5 percent per year. Envision a half-decade transition. During the half decade, our aim is to keep output growing about 2.5 percent per year while gradually raising the share of output that consists of investment goods (from 20 percent to 24 percent), while gradually reducing the share that consists of consumer goods (from 80 percent to 76 percent). This will happen if consumer goods production grows about 1.5 percent per year while investment goods production grows a little over 6 percent per year.

Here's the arithmetic. Suppose that in year 0 output is $1,000,

consumption $800, and investment $200. If consumption grows 1.5 percent per year for five years, in year 5 it will be $800 × (1.015)^5 = $862. If investment grows 6.2 percent per year for five years, in year 5 it will be $200 × (1.062)^5 = $270. So output in year 5 will be $862 + $270 = $1,132; hence output will have grown approximately 2.5 percent per year, because $1,000 × (1.025)^5 = $1,131. But now consumption will be 76 percent of output ($862/$1,131 = 0.76) and investment will be 24 percent of output ($270/$1,131 = 0.24). From then on, we envision the shares (76 percent, 24 percent) remaining constant so that output, consumption, and investment all grow at the same rate—a bit higher than 2.5 percent per year for many years due to the greater investment rate (24 percent vs. 20 percent). Thus, in a more realistic model with labor force growth and technological progress, during the transition to a permanently higher saving rate, consumption might grow 1.5 percent per year instead of its normal 2.5 percent per year. This below-normal growth still implies a sacrifice, but consumption does not literally decline.

The Open Economy's Long-Run Response to an Increase in the Saving Rate

Thus far we have assumed a *closed* economy. But what happens if the economy is *open* to foreign borrowing or lending, to exports and imports?

To answer this question, economists have constructed a model. In order to isolate the impact of a difference in the saving rate, the model assumes that our economy and the economy of the rest of the world are identical except that saving rates differ. There's no point keeping you in suspense. Here's what the model tells us. If our saving rate exceeds the rest of the world's, then we will eventually achieve higher wealth, income, and consumption per person than the rest of the world. We will be a creditor nation and receive net interest payments from the rest of the world. We will consume more by being a net importer of goods, running a trade (in goods) deficit. But our current account will be in surplus, because our net interest earnings will exceed our trade deficit.

Conversely, if our saving rate is less than the rest of the world's, then we will eventually suffer lower wealth, income, and consump-

tion per person than the rest of the world. We will be a debtor nation and make net interest payments to the rest of the world. We will consume less by being a net exporter of goods, running a trade (in goods) surplus. But our current account will be in deficit, because our net interest payments will exceed our trade surplus. So according to economic analysis, our *relative* saving rate—how our saving rate compares with that of the rest of the world—is one key determinant of our current account and trade balances. And our major conclusion remains unchanged in an open economy with trade and capital flows: A permanent increase in our saving rate will eventually achieve higher consumption per person.

But does the economist's model capture all the complexity of the world economy? Of course not. Is the relative saving rate really the only determinant of a nation's relative standard of living and its current account and trade balances? Of course not. Don't things like natural resource endowments, real investment opportunities, and entrepreneurship matter? Certainly they do. And aren't private property rights and free markets important? Yes, they are. Welcome to the real world. It's complicated.

But in a complicated world, we've got to simplify to make any progress. Not only that, we've got to concentrate on issues we can do something about. We can't change our basic natural resource endowment. But we can do something about our national saving rate. So that's where we focus our model and our attention. And a simple message emerges: In an open economy, as well as a closed economy, a high relative saving rate is a key factor in obtaining a high relative standard of living.

In a world of international competition, saving is a key factor. We should concentrate on achieving a high relative saving rate, thereby eventually accumulating more wealth per person than other nations. Which nation will emerge with a standard of living that is second to none? Almost surely it will be the nation that sustains the highest saving rate decade after decade.

I'm sure you want to know some of the assumptions behind our model. Let me begin by telling you its simplifications. The model has two "countries"—our country and the rest of the world (all other countries consolidated). Our country is small relative to the rest of the world, so that its behavior has little impact on the values of the

rest of the world's economic variables. Except for scale, our country's economy and the world's economy are identical in every respect you can think of: they have the same population growth rate, the same production technology, and so on. And initially they have the same saving rate.

Now comes a key simplifying assumption: perfect capital mobility. This means that if savers discover they can earn a higher interest rate by lending abroad rather than at home, they shift their funds. This implies that capital will flow until the interest rate on domestic investment equals the interest rate on world investment.

But since the interest rate depends on the physical productivity of capital, and this in turn depends on capital per worker, it follows that our country and the rest of the world will always have the same capital per worker and hence the same domestic output per worker. If our capital per worker stays the same as the rest of the world's as both labor forces grow, then our domestic physical investment per worker must be the same as the rest of the world's.

I'm sure you won't be surprised to learn that with everything identical, including the saving rate, these assumptions imply that our current account and trade balances are both zero, and that the United States is neither a creditor nor a debtor nation. It turns out that our domestic absorption—consumption plus investment—exactly equals our domestic output, so that any imports are exactly matched by exports, and our trade balance is zero.

Also, our country's domestic investment exactly matches our own saving, so any interest payments our savers earn from abroad are matched by interest payments from our firms to the world's savers; thus, *net* interest payments are zero. So our current account balance, which includes payments for goods and interest, is also zero. Our wealth is equal to our domestic capital stock. We own the same amount of the rest of the world's capital stock as the rest of the world owns of our capital stock. So we are neither a creditor nor debtor nation.

But now suppose our country permanently raises its saving rate, so that our saving rate exceeds the rest of the world's. What happens? The moment our saving rate rises, our consumption rate falls. So our domestic absorption—consumption plus investment—will immediately fall below our domestic output. Thus, the excess of output over absorption will be exported, and the immediate result will be a trade

surplus. With net interest payments still zero, the result will also be a current account surplus equal to the trade surplus.

But net interest payments will not remain zero. Our saving now exceeds our domestic investment. Rather than confront diminishing returns at home, our excess saving will flow abroad to finance real investment in the world economy. But this means that our savers will earn net interest payments from abroad.

As our savers accumulate wealth—claims on the world capital stock—wealth per worker and income (including interest income) per worker will rise. But this means that consumption per worker will also rise, after its initial setback. It turns out that as long as the interest rate exceeds the growth rate of labor, consumption per worker will eventually surpass its initial level. When this happens, domestic absorption (consumption plus investment) will exceed domestic output, and the trade balance will reverse. We will become a net importer of goods.

In the final steady state, we will have a current account surplus, but a trade (in goods) deficit. The net interest payments our savers earn from the rest of the world will exceed the net payments we make to buy goods. We will also be a creditor nation, owning more of the rest of the world's capital stock than it owns of our capital stock.

Note that in the short run, after we raise our saving rate, we initially run a trade (in goods) surplus. The surplus gradually becomes a deficit only after we accumulate wealth, become a creditor nation, and earn enough interest from abroad to finance higher consumption per worker than the rest of the world.

8

Growth Through Tax Reform

How can we promote economic growth? In chapter 7, we saw that one way to promote growth is fiscal discipline: the government must limit its borrowing (except in a recession), so that businesses can borrow virtually all the funds that savers make available, and can use these funds to finance investment. In this chapter, we examine another way to promote growth: tax reform. We can state our prescription simply: Instead of taxing income, tax consumption.

Our Income Tax

To understand our proposal, let's first see how our income tax system works. Once a year, on or before April 15, each household "files" its annual income tax "return"—that is, the household fills out a form (its *return*), often with the help of an accountant, reporting all the income received in the previous calendar year; then it figures the tax owed and mails its return to the Internal Revenue Service of the U.S. Treasury. You're allowed to file before April 15, but many people, like many students with their term papers, wait until the last day.

But does the government really wait until April 15 to get its money, trusting each household to save up enough all year to pay its tax? Of course not. The government requires employers to take money out of each paycheck—this is called *withholding*—and send it to the gov-

ernment. Also, some households are required to make four quarterly payments to the government for the tax they think they will owe on investment income (interest, dividends, and capital gains). So by the time the household files its annual tax return, the government has already collected a lot of tax. Suppose household X figures out that it owes $10,000 in tax, but the government has already collected $8,000. When it mails in its tax return, it encloses a check for $2,000. What if the government has already collected $12,000? Is household X out of luck? No. When it mails in its tax return, it requests a $2,000 refund. A month or two later, it will receive a nice check from the U.S. Treasury for $2,000. So it is a mistake to think that income tax is paid on April 15. It's paid all year. What happens on April 15 is that the household files its tax return: it encloses a check if it owes more for the year than has already been collected, but it claims a refund if the reverse is true.

The federal income tax is *progressive*. This means that the higher a household's income, the greater the *percentage* it pays in tax. If instead the percentage stayed the same, it would be called a *proportional* (or *flat*) tax; and if the percentage went down, it would be called a *regressive* tax. A household doesn't owe any tax until its annual income exceeds a certain amount; only income above that amount is *taxable income*. Here's an example that uses round numbers that were approximately correct in the year 2003.

Suppose in 2003 your household received a nice income (from salaries and interest) of $400,000, and your family has four members (husband, wife, and two children). For each member there is a *personal exemption* of $3,000, for a total of $12,000. You are also allowed to take a set of *itemized deductions*—for example, for mortgage interest payments on your home, state and local property taxes, and charitable contributions—and these total $38,000. Then your *taxable income* would be $350,000 ($400,000 − $12,000 − $38,000). Your taxable income would be taxed according to Table 8.1.

Your first $14,000 would be taxed 10 percent, your next $43,000 (from $14,000 to $57,000) would be taxed 15 percent, your next $58,000 (from $57,000 to $115,000) would be taxed 25 percent, and so on. The table shows how you would compute your tax; adding the tax on each bracket of income gives a total of $97,660. Note that

Table 8.1

How to Compute Your Income Tax Under the 2003 Income Tax Schedule

Here are the tax brackets and tax rates:

Taxable income ($)	Bracket tax rate (%)
0–14,000	10
14,000–57,000	15
57,000–115,000	25
115,000–175,000	28
175,000–312,000	33
312,000 and over	35

If your *taxable* income is $350,000, what is the total tax that you must pay? You need to add six numbers:

.10	x	$ 14,000	=	$ 1,400
.15	x	$ 43,000	=	$ 6,450
.25	x	$ 58,000	=	$14,500
.28	x	$ 60,000	=	$16,800
.33	x	$137,000	=	$45,210
.35	x	$ 38,000	=	$13,300
Total Tax			=	$97,660

Your total tax as a percent of your taxable income is 28 percent ($97,660/ $350,000), even though you are in the 35 percent tax bracket (the last $100 you earn is taxed $35).

$97,660/$350,000 = 28 percent, so your tax is 28 percent of your taxable income, even though your last dollar earned is taxed 35 percent. We say you're in the 35 percent "tax bracket"—your last dollar of income is taxed 35 percent—but your tax is only 28 percent of your taxable income. Also, it's important to note that if any of your income consisted of *capital gains* (for example, if you bought shares of corporate stock five years ago for $10,000, and this year sold the shares for $15,000, your capital gain would be $5,000) or *dividends* (if you own shares of stock in companies that pay dividends to shareholders), these kinds of income would be taxed separately at a rate of only 15 percent (even though you are in a 35 percent tax bracket for all other kinds of income).

Does our income tax discourage economic growth? It does, but not because tax rates are too high. If our top income tax rate were twice as high—70 percent instead of 35 percent—there would be cause for concern. With a 70 percent rate, a highly-paid person might reason: "If I work more and earn another $1,000, I will only get to keep $300; maybe it's not worth the effort." Our top rate was 70 percent in 1980. But the tax cut of 1981 reduced the top rate to 50 percent, the tax reform act of 1986 reduced it to 28 percent; and though the tax act of 1993 raised the top rate to 39.6 percent, the economy boomed in the rest of the 1990s, with hard work and entrepreneurial effort especially evident in the high-technology sector (it should be noted that the top tax rate on capital gains was reduced to 20 percent in the 1990s, and this probably helped motivate entrepreneurs). Clearly, most people were willing to work hard in the 1990s despite a top tax rate of 39.6 percent. Moreover, the United States is a low-tax country by international standards: tax revenue is about 30 percent of GDP (20 percent federal, 10 percent state and local); in western Europe, it's 40 percent, and in Scandinavia (Denmark, Sweden, Norway, and Finland), it's 50 percent. Yet the highly taxed Scandinavian economies have produced as much GDP per person as the United States. So while we would like lower tax rates (and lower prices, too), most people are willing to work hard at our current tax rates.

Our income tax discourages economic growth, not because tax rates are too high, but because it discourages saving. Under an income tax, you pay the same tax whether you consume or save most of your income this year. And the more you save this year, the more you will be taxed in the future on the interest, dividends, or capital gains you earn from this year's saving. Alas, the income tax disfavors saving. But saving and investment are the engines of economic growth. So the question arises: Can't something be done about this discouragement to economic growth?

Tax Consumption Instead of Income

Many people are gripped by a fear of heresy when they hear this proposal, as if on the sixth day God had said, "Let there be an income tax." True, the income tax has been the centerpiece of the U.S. tax structure for several decades, and the propriety and wisdom of taxing income have come to be taken for granted.

Yet many economists have long advocated taxing households on their consumption, not income. In 1995, senators Pete Domenici (R-NM), Sam Nunn (D-GA), and Bob Kerrey (D-NE) introduced a bill that would have established the Unlimited Savings Allowance Tax (USA Tax). The USA Tax would have converted the personal income tax to a personal consumption tax by making all saving tax deductible. The crucial difference between a personal consumption tax and an income tax is simply this: Under a consumption tax, saving would be *tax deductible*; every hundred dollars saved would be a hundred dollars that is exempt from tax.

Does a Personal Consumption Tax Favor the Affluent?

The most common reaction to the proposal is, "It favors the affluent, who can most afford to save." But that reaction is based on a fundamental misunderstanding. Why?

The source of the reaction is the mistaken assumption that the tax rates in the tax tables will be unchanged when saving is made tax deductible. If these tax rates were unchanged, then deductibility would indeed favor the affluent, who can most afford to save.

But, who says the tax rates must stay the same? In fact, if the rates were unchanged, then less total revenue would be collected due to the saving deduction, and our budget deficit would get even larger. So, when we convert to a personal consumption tax and make saving tax deductible, the rates in the tax table must be raised to keep tax revenue constant. Please note that these rate increases would not raise the dollar tax payment of the average household. They would simply keep the average dollar tax payment the same despite the new saving deduction.

But how should the rate increases be apportioned among income classes? Suppose Congress wants each income class to pay the same revenue it paid under the income tax. Since high-income households save most, their tax rate must be raised most. And since low-income households save least, their tax rate must be raised least.

So we are advocating conversion of the income tax to an *equally progressive* consumption tax. How do we achieve it? Divide the population into income classes. Calculate the total tax revenue paid by

each class under the current income tax. Then set the new tax rates under the consumption tax so that each income class pays roughly the same total tax as before.

Citizens disagree about how the nation's tax burden should be distributed across income classes; they disagree about how *progressive* the tax system should be. The crucial point to grasp is that making saving tax deductible by converting from an income to a personal consumption tax is neutral with respect to distribution. How the rates are set, under either an income or a consumption tax, determines how the burden is distributed across classes. The choice of tax base— income versus consumption—is completely separate from the choice of distribution—how to set tax rates for each class. Thus, conversion to a personal consumption tax has been advocated by economists who disagree about distribution but agree about the need to raise our national saving rate. So if someone is either for or against conversion because he thinks it favors the affluent, then his reaction is based on a misunderstanding. Congress can make the new consumption tax have more, less, or the same progressivity as the current income tax, simply by adjusting the tax rates in the tax table.

The IRA Principle

Making saving tax deductible may sound like a radical departure, but it isn't. We've already taken several initial steps in that direction under our income tax.

In the early 1980s, the tax law was amended to allow Individual Retirement Account (IRA) saving to be tax deductible. Unfortunately, the IRA has two key restrictions. First, there is a ceiling on the amount of annual saving that is tax deductible. Second, there is a penalty for withdrawal before retirement. As its name suggests, the purpose of the IRA is to encourage a limited amount of saving for retirement.

Under a personal consumption tax, all saving for any purpose would be tax deductible. There would be no limit on the tax-deductible amount. When funds are withdrawn to finance consumption, the consumption would be taxed. But there would be no penalty for withdrawal per se.

The IRA is not the only step we've taken toward a personal consumption tax. Under current tax law, if a person's employer contrib-

utes $1,000 to his pension fund, the employee does not pay tax on this $1,000 of income. Because $1,000 of his income has been channeled into saving, it is tax deductible for the employee. Once again, there is an important restriction: The saving must be for retirement.

These restrictions make sense if the aim is to encourage only provision for retirement. But if the aim is to keep our future standard of living second to none, then all saving warrants encouragement. To promote this objective, we must implement the IRA principle more thoroughly by making all saving tax deductible, thereby converting our income tax to a personal consumption tax.

Practical Features of a Personal Consumption Tax

But isn't it impractical to ask everyone to keep receipts for everything they buy? The breakthrough came when some practical person realized that we don't need to add a huge number of purchase receipts to figure out a household's consumption. This ingenious person realized that all we have to do is "follow the cash." We can determine a household's consumption by adding and subtracting only a few items.

The basic insight couldn't be simpler. Almost all consumption is financed by money or checks, which I will call "cash" (perhaps with a delay made possible by a credit card). We simply follow the cash. Add the cash inflows. Subtract the nonconsumption cash outflows. The remainder must have been used for consumption, so tax it. A personal consumption tax return is illustrated in the appendix to this chapter.

Here's a simplified example. Suppose a household earns $60,000 in salaries, receives $4,000 in interest and dividends, and sells stocks and bonds for $6,000, for a total cash inflow of $70,000. If the household increases its saving account balance by $8,000, and buys new stocks and bonds for $12,000, its total nonconsumption cash outflow is $20,000. Therefore, its consumption is $50,000 ($70,000 minus $20,000).

I know what you're thinking. What about the treatment of housing and other consumer durables? What about the treatment of gifts and bequests? What about the borderline between consumption and saving? I won't kid you. A consumption tax has some practical problems. They have been examined in detail by experts. Remember, I

didn't claim that a consumption tax would be simpler than an income tax, just that it wouldn't be more complicated. I don't advocate a consumption tax for simplicity, but to protect the future standard of living. Here are some possible solutions to these practical problems.

Economists agree that when you purchase a consumer durable, like a car or a house, you are making an investment. Then you consume the services of the durable over many years. So one option is this: If you borrow to finance the durable, the loan can be excluded from cash inflow, so only the down payment is taxed in the year of purchase. But in each subsequent year, you will not be allowed to deduct the loan repayment. Thus, you will be taxed each year on the loan repayment, which is a rough approximation of your consumption that year.

When a donor gives a gift or bequest, she may get pleasure out of it. For that matter, a saver may get pleasure out of saving. But like the saver, the donor is not consuming; she is not drawing resources—land, labor, and capital—away from real investment. Her abstaining from consumption helps our future standard of living. If you favor a consumption tax to raise the future standard of living, then you should agree that the gift or bequest should be treated as tax deductible to the donor.

What about the donee—the recipient? The gift or bequest is a cash inflow. If he saves it, then he obtains an equal deduction, so the gift or bequest is not taxed. If he ever consumes it, he will be taxed in that year. If he never consumes it, then it will never be taxed. This makes sense, because as long as he abstains from consumption, he helps advance the future standard of living.

Now, some favor a consumption/gift/bequest (CGB) tax because they want the donor to be taxed. They object that under a consumption tax, gift givers will permanently escape tax, and that this is unfair. They object that donees will escape tax until they actually consume, and that this is unfair. Why is it unfair? Because, they argue, donors and donees may get pleasure or security from gifts and bequests.

But I have a different perspective. My top priority is the future standard of living. I want to reward our citizens when they help raise it. As long as they do not consume, the donor and donee are helping by releasing resources for investment. So they should not be taxed.

Incidentally, for the same reason, it makes sense to abolish all estate and gift taxes and make up the lost revenue by raising high-bracket consumption tax rates. The affluent who transfer wealth, rather than consume it, are releasing resources for investment. Abolition of estate and gift taxes, together with conversion to a personal consumption tax, gives the affluent an incentive to preserve wealth and refrain from consumption, exactly what's needed to raise everyone's future standard of living.

Finally, what about the border between consumption and investment? Let me examine one item: education. I know this will sound self-serving, coming from a professor. But the fact is that for at least two hundred years, economists have emphasized that education is an investment in human capital. Like machinery, education raises the productivity of workers. Sure, college can be enjoyable—it is partly consumption. But the investment component of a household's expenditure for education or training should be treated as tax-deductible investment under a consumption tax. Perhaps a simple rule of permitting 50 percent deductibility, up to some ceiling, would be a reasonable treatment.

Consider what this means: When you set money aside for college tuition while your child is in diapers, it is, of course, tax-deductible saving. But even when you withdraw the funds to pay tuition, half (up to a ceiling) would remain tax deductible, because an expenditure on education is partly investment.

Despite these possible solutions, there is no denying that the consumption tax has some thorny practical problems. On the other hand, a consumption tax is simpler than an income tax in certain respects. Employee compensation is complex under the income tax: Should stock options be taxed as ordinary income or be given special capital gains treatment? But compensation is simple under the consumption tax: Only cash received by the household is counted in its cash inflows. A capital gain is complex under the income tax: Should an attempt be made to make up for the advantage of deferring tax until the year of sale? But a capital gain is simple under the consumption tax: Only cash from the sale of stock is counted in its cash inflows. Finally, saving is complex under the income tax: Which savings vehicles (IRA, 401(k) plan) should be granted a tax deduction? But saving is simple under the consumption tax: All saving is tax deductible.

Although our focus is on the household tax, a brief comment about the business tax is warranted. A corporate income tax does not fit with a personal consumption tax. The USA Tax proposes replacing the corporate income tax with a subtraction value-added tax (VAT). Under a VAT, a business is taxed on its sales revenue minus its purchases from other firms. A VAT is a consumption tax because each business can subtract the purchase of capital goods—investment—in the year it occurs, so that firms are taxed on output (value added) minus investment, which equals consumption. Thus, the VAT makes investment tax deductible just as the personal consumption tax makes saving tax deductible. The two taxes fit together: They encourage saving and investment.

The Horizontal Redistribution Effect

Conversion to an equally progressive consumption tax will raise national saving. To help make the point as clear as possible, two affluent persons, each with $500,000 of income, have agreed to be extremists. Person S has agreed to save everything and consume nothing, while person C has agreed to consume everything and save nothing.

Of course, emaciated S can only keep this up long enough for readers to grasp the pedagogical point. Rather than ridicule S and C for extremism, be grateful for their voluntary service in the cause of clarity. To compensate them for their service, let's flatter their egos by letting them be the only two people in the affluent $500,000 income class.

Table 8.2 presents our example. Under an income tax of 20 percent, S and C each pay $100,000 in tax, so total tax revenue is $200,000. Person S saves $400,000 and C nothing, so total saving is $400,000. Person C consumes $400,000 and S nothing, so total consumption is $400,000.

Now Congress makes saving tax deductible and converts the income tax to a personal consumption tax. What rate must Congress set for the affluent class in order to raise the same total tax revenue, $200,000? Because S and C have graciously agreed to be extremists, the answer is easy. Congressional eyes gaze insidiously on C, who alone will now pay tax. Emaciated S, though weak from lack of nourishment, manages a proud sneer, whispering that she is now exempt

Table 8.2

Conversion to a Consumption Tax Raises Saving

20% Income tax	Income ($)	Tax ($)	Consumption ($)	Saving ($)
Person C	500,000	100,000	400,000	0
Person S ·	500,000	100,000	0	400,000
TOTAL	1,000,000	200,000	400,000	400,000
40% Consumption tax	Income	Tax	Consumption	Saving
Person C	500,000	200,000	300,000	0
Person S	500,000	0	0	500,000
Total	1,000,000	200,000	300,000	500,000

from tax. Since C must pay $200,000 in tax, a 40 percent consumption tax rate will do the trick because C takes no saving deduction from her cash inflow of $500,000.

What is the impact of tax conversion? Total tax revenue collected from the affluent remains $200,000. But now, instead of $100,000 coming from each, all $200,000 comes from C. Person S, believe it or not, now saves all $500,000 of her income, instead of "only" $400,000, so total saving rises by $100,000, to $500,000. And since C is forced to cut her consumption due to her higher tax, total consumption falls by $100,000 to $300,000.

What has really happened is this: Tax conversion has caused $100,000 of cash to be redistributed from C to S, because C's cash falls by $100,000 (from $400,000 to $300,000) and S's rises by $100,000 (from $400,000 to $500,000). Person C would have consumed the $100,000, but person S saves it. Hence, total consumption falls $100,000, and total saving rises $100,000.

I call this increase in total saving due to the shifting of cash among the affluent the *horizontal redistribution effect*. It is "horizontal" because it is a shift among the affluent, not across different income classes.

Note that the horizontal redistribution effect has nothing to do with the incentive to save. Our two gracious volunteers have agreed to be impervious to incentives. They doggedly stick to their extremist behavior even when saving is made tax deductible. Person C continues to save nothing, despite the deductibility of saving. And emaciated S

would no doubt save more if she could, but she can't, because she's already saving her entire income.

Through the horizontal redistribution effect, making saving tax deductible will raise total saving, even if no one responds to the greater incentive to save. This fundamental point is overlooked in the public debate, even by some sophisticated analysts. The debate over making saving tax deductible usually runs as follows. Advocates claim that the average individual will raise his propensity to save—the fraction of his after-tax income that he saves—if saving is deductible. Opponents deny it. Both sides assume that total saving will rise only if the typical individual raises his propensity to save.

But our example shows that this is not so. Extremists S and C did not change their propensity to save: S kept it at 100 percent, and C kept it at 0 percent. Each was completely unresponsive to the new incentive to save. But total saving rose through the horizontal redistribution effect. Why? Because deductibility shifted cash away from C, who would have consumed it, to S, who saved it.

The horizontal redistribution effect, of course, applies not only to the affluent, but to every income class. Congress can adjust the tax rate of each income class so that the consumption tax raises roughly the same total revenue from that class as did the income tax. But within each class, there will be a horizontal redistribution effect. Above-average savers will enjoy a tax cut, and below-average savers will suffer a tax increase. Cash will shift horizontally from persons with a relatively low propensity to save to persons with a relatively high propensity to save. So total saving will increase.

Although the horizontal redistribution effect applies to all classes, its impact is most important among the affluent. Not surprisingly, there is much more variation in the propensity to save among the affluent than among low-income households. Most low-income households save very little. But among the affluent, some households save a large fraction of their income, and others dissave— consuming more than their income. For example, in a study done by researchers at the Federal Reserve, roughly 20 percent of affluent households saved two-thirds of their income, another 20 percent saved roughly half, but 20 percent dissaved (consumed more than their income). My colleague Ken Lewis and I used the Federal Reserve data to estimate how much the horizontal redistribution

effect would raise total household saving. We estimated an increase of roughly 10 percent.

We can now see whose taxes rise and whose fall by making saving tax deductible and adjusting tax rates to keep the new consumption tax as progressive as the income tax it replaces. Deductibility does not favor one income class over another. Instead, within each income class, above-average savers enjoy a tax cut and below-average savers suffer a tax increase; the average saver in the income class pays the same tax.

The Incentive Effect

Extremist C continued to consume all her after-tax income, despite the deductibility of saving. Her consumption fell only because she had less cash. But wouldn't you take advantage of the new deduction by saving a little more? Wouldn't you respond to the new incentive to save? Wouldn't the average person?

How would citizens react to the headline "Saving Now Tax Deductible"? Word would spread among ordinary taxpayers, not merely shrewd tax planners, that every hundred dollars saved is a hundred dollars not taxed. Many citizens ask: "How can I reduce my taxes?" Now there would be a clear answer: "Save." The subheadline would read: "No Restrictions." The saving can be for any purpose. It can be withdrawn without special penalty. And there is no limit to how much saving is deductible.

Note the difference between open-ended, unrestricted deductibility and IRAs. With IRAs, a person must worry about whether he is over-saving for retirement. "Suppose I need cash in five years? I'll regret that it's tied up in my IRA." But with unrestricted deductibility, the person must think no further than this year. "If I can get through this year, let me save. I can always withdraw it next year, or in five years, for any purpose, without penalty." Soon, however, a citizen will overcome his enthusiasm with this year's tax saving and face what happens in the future when he withdraws funds to finance consumption. How will the typical citizen react when he hears the IRS say: "We may not get you now, but we'll get you later"?

Some citizens may succumb in despair. "Why save when it only postpones the tax?" some will ask. But others will react: "Postpone-

ment is still worth it. After all, if I'm lucky, the world may end in the meantime." Others, however, will raise this question: "What if I cross the finish line of life never having withdrawn my savings for consumption? What if I leave a large bequest to my heirs at death?"

Good question. Some tax analysts want a bequest at death to be taxed. They worry about letting misers sneak across the finish line of life without paying their fair share. They advocate a consumption/gift/bequest (CGB) tax instead of a consumption tax. But I'm less worried about thrifty people than I am about our nation's future standard of living. The fact is that only actual consumption, not bequests, draws resources away from investment and reduces our future standard of living. So I side with other consumption tax advocates who say: "If you cross the finish line without consuming, you win; a bequest is not consumption."

While it's only a guess, I suspect that taxing only actual consumption, not bequests, will raise our national saving rate. I have a hunch that the thought of permanently beating the IRS would inspire quite a few citizens to save more. When these citizens hear someone say, "Why save? They'll only get you later when you spend it," these citizens will reply with joy, "They'll never get me, because I'll never spend it; I'll wave it at the IRS from my deathbed and pass it on to my children and grandchildren."

The Postponement Effect

When taxes are compared, attention focuses on incentives. But there is another crucial aspect of taxes that is usually overlooked: Which tax most postpones collection over an individual's "life cycle"?

Perhaps the most important property of the consumption tax is this: it is "the great postponer." For this reason, perhaps more than any other, a consumption tax should result in a greater accumulation of capital in the economy than an income tax.

Imagine that each individual is a "life cycler." A life cycler plans ahead. He recognizes that someday he will retire, and upon retirement, alas, his income will fall further than his desired consumption. While he works, he must save, so that he can dissave in retirement to finance his consumption. Of course, not everyone is a life cycler, but enough people plan ahead, however imperfectly, to make the life cycler worth studying.

A life cycler's consumption is less than his income during most of his work life, and greater than his income during retirement. Thus, when the income tax is converted to a consumption tax, the young person just beginning his life cycle will enjoy a tax cut during his work life and will incur a tax increase in retirement. Conversion causes postponement of some of his lifetime tax.

But this means that conversion to a consumption tax increases the ability of workers to save and accumulate wealth. Tax postponement therefore results in greater accumulation of wealth by the typical life cycler and hence a greater capital stock for the economy.

Thus, a consumption tax should achieve a higher capital stock than an income tax for three reasons: the horizontal redistribution effect, the incentive effect, and the postponement effect.

Is a Consumption Tax Equivalent to a Labor Income Tax?

Instead of a saving deduction, why not enact a capital income exemption? Under this alternative tax reform, you wouldn't get a deduction when you save. But when you earn interest, dividends, or capital gains, your capital income would be exempt from tax. Like a saving deduction, a capital income exemption gives an incentive to save.

A saving deduction converts the income tax into a consumption tax. A capital income exemption converts the income tax into a labor income tax. Both conversions give an incentive to save. For this reason, some analysts have claimed that a consumption tax is really equivalent to a labor income tax.

But it isn't. And it is crucial to understand why.

While the consumption tax is "the great postponer," the labor income tax is "the great upfronter." Under a labor income tax, the government raises all tax revenue from workers and none from retirees. So a life cycler pays all his tax "up front"—during the work stage of life. Compared to a consumption tax, a labor income tax imposes a greater tax burden on the worker, thereby reducing his ability to save. Hence, a labor income tax would achieve a smaller capital stock than a consumption tax.

In fact, the labor income tax is more of an upfronter than the regu-

lar income tax, which at least postpones some tax to retirement. So it is even possible that, despite its incentive to save, a labor income tax would achieve a lower capital stock than a regular income tax. By contrast, a consumption tax gives an incentive to save, and is also a greater postponer than an income tax, so it will clearly achieve a higher capital stock for the economy. The labor income tax gives the worker an incentive to save, but reduces his ability to do so. The consumption tax gives the worker an incentive to save, and improves his ability to do so. Since the consumption tax should result in a significantly larger capital stock than the labor income tax, it makes no sense to call them equivalent.

The two taxes also differ fundamentally with respect to fairness. The best way to communicate this difference is to shine the spotlight on a notorious character: the lazy heir. For some, the only worth of this brazen individual is his pedagogical value. For others, he is a source of secret admiration. At any rate, who is he? The lazy heir inherits a large fortune, uses it to finance a high level of consumption, never works a day in his life, and dies leaving nothing to his own children, because, he says, he doesn't wish to spoil them.

Now, what tax would the lazy heir owe under a labor income tax? Zero. At the annual April 15 news conference at his plush estate, the lazy heir holds up his tax return—an empty sheet of paper. With servants surrounding him, he complains that it is most fortunate that he owes no tax, not having worked, because he needs every bit of his fortune to maintain his mansion. Needless to say, he is a favorite on the evening news.

Under a consumption tax, however, the plight of our lazy heir would be severe. His high consumption would incur a high tax. As he sells the stocks and bonds he inherited, his cash inflow would record the sale of assets. Since there is no corresponding deduction, his taxable consumption would match his asset sales.

In fact, a consumption tax would tax the lazy heir more heavily than an income tax. Under an income tax, he would be taxed on capital income. But under a consumption tax, he would be taxed on the wealth he "decumulates" each year. Suppose the lazy heir, for spite, converts his fortune to cash, and places it under his luxurious pillow so that it earns no income. He would owe no income tax. But he would still owe substantial consumption tax.

Thus, with respect to fairness, a consumption tax is surely not equivalent to a labor income tax. Just ask the lazy heir.

How to Gradually Phase in a Personal Consumption Tax

Conversion to a consumption tax must be phased in gradually for two distinct reasons: to ensure a smooth macroeconomic transition, and to avoid double taxation. Let's consider each in turn.

First, macroeconomics. Recall what we are trying to do. We want to gradually raise the percentage of our national economic pie that goes to investment. This requires that we gradually reduce the percentage that goes to consumption. But our pie per person grows in a typical year. So it is possible to gradually reduce the percentage that goes to consumption without ever reducing the absolute size of the consumption slice per person. This should be our aim: to gradually reduce the percentage without ever reducing the absolute amount per person.

We want to slow the growth of consumer goods, so that workers who quit or retire in the consumer goods sector are not replaced, and most new jobs open up in the rapidly growing investment goods sector. But we do not want to make the output of consumer goods literally decline, because this would result in involuntary layoffs and a transitional recession. No one knows how much our population would cut its consumption demand if the entire population were converted to the consumption tax in a single year. It would therefore be completely irresponsible to convert the whole population at one time.

Now let's consider the double taxation problem. Under an income tax, a person pays tax on the income he saves, but no tax on his retirement consumption. Under a consumption tax, a person pays no tax on the income he saves, but pays tax on his retirement consumption. Now consider the "lucky" person who gets caught in the transition. He has paid tax on the income he saved. And he will also pay tax on his retirement consumption. He has a complaint: "I've been double taxed."

Here's how to handle both problems. First, for macroeconomic smoothness, the population should be converted to the consumption tax in stages. Perhaps the best method would be to phase in a cross section of the population each year over a half decade. Once you convert to the consumption tax, you convert for life. Cross-section

phasing has an important advantage: because a similar mix of ages and incomes converts each year, there should be minimal disruption to the sales of particular industries.

Second, to avoid double taxation, each household would compute its "deductible old wealth" on the first day of the year it converts to the consumption tax. Its deductible old wealth would consist of wealth that would not be taxed again under the income tax, but would be taxed under the consumption tax without special protection. Wealth likely to be given as a gift or bequest is already protected because the donor would not be taxed under the consumption tax. The household would be permitted to deduct a percentage of its deductible old wealth each year over half a decade.

The Personal Consumption Tax Versus Other Consumption Taxes

The other consumption taxes are the retail sales tax, the value added tax (VAT), and the "flat tax." Each is simpler than the personal consumption tax, and each will raise the national saving rate. So why not abolish the income tax and replace it with one of these consumption taxes?

There's one little problem with this strategy. The current income tax is progressive: Affluent households pay a higher ratio of tax to income than nonaffluent households. Replacing the income tax with either a retail sales tax, a VAT, or a flat tax would significantly shift the tax burden away from the affluent to the nonaffluent. For example, in 1995 the Office of Tax Analysis of the U.S. Treasury estimated that replacing the income tax with a retail sales tax or VAT would reduce the federal tax burden of the richest 5 percent by 39 percent, and of the richest 1 percent by 55 percent. Replacement by a flat tax—where households pay a flat rate of about 20 percent above a $30,000 exemption—would reduce the federal tax burden of the richest 5 percent by 21 percent, and of the richest 1 percent by 36 percent—thanks to the exemption.

Now, citizens can disagree about whether this shift in the tax burden from affluent to nonaffluent households is desirable or undesirable, fair or unfair. But there is no disagreement about one thing: A proposal to replace the income tax by any of these other consumption taxes will provoke a politically divisive debate between the affluent and the nonaffluent. The debate over how the tax burden should be distributed

will dominate the main objective: raising the national saving rate.

This is why moderate Republican Domenici and moderate Democrats Nunn and Kerrey decided to support replacing the income tax by an equally progressive personal consumption tax. Their USA Tax bill, introduced into the U.S. Senate in 1995, would have set tax rates in its tax table to achieve roughly the same distribution of the tax burden as the income tax it would have replaced. The strategy behind the USA Tax was simple: Let's first pass a fundamental tax reform that will raise the national saving rate without shifting the tax burden from affluent to nonaffluent. Once we have a personal consumption tax, we are always free to debate whether to adjust the rates in the tax table, thereby shifting the distribution of the tax burden in one direction or another. Unfortunately, the debate over fundamental tax reform deadlocked in the past decade because of the false impression that any consumption tax must shift the tax burden from the affluent to the nonaffluent. The USA Tax bill shows that a consumption tax can be just as progressive as our current income tax. Hopefully, the USA Tax will be given consideration the next time the nation debates fundamental tax reform.

Fairness: The Personal Consumption Tax Versus the Personal Income Tax

Is it fairer to tax a person according to her consumption instead of her income? The income tax advocate thinks it isn't, asserting: "Income is a better measure of ability to pay, and persons should be taxed according to their ability to pay. Consider the person with high income but low consumption. She should pay a high tax, because her ability to pay is high, not a low tax simply because she chooses to save a lot and consume a little."

The consumption tax advocate replies: "It's fairer to tax a person according to what she takes out of the economic pie rather than according to what she contributes to it. When a person produces and earns income, a contribution is made to the pool of available goods and services. Production potentially adds, rather than subtracts, from others' economic well-being. But when people actually withdraw resources for their own consumption, then these resources are not available for others to consume or businesses to invest in plant, equipment, and technology, thereby raising everyone's productivity and earnings in the future."

"But what about ability to pay?" asks the income tax advocate.

"The ability-to-pay principle," replies the consumption tax advocate, "has an element of expediency: Tax a person more, simply because the person is able to pay more. But a principle of fairness ought to consider how a person's economic behavior affects others. From this perspective, it is fairer to tax a person according to what that person subtracts from, rather than adds to, the economic pie.

"Consider Connie and Sally. They have the same production and income, but Connie uses her entire income to consume goods and services for her own enjoyment, while Sally saves a large share of her income and uses only a small share to withdraw consumption goods, leaving resources for others to consume and invest. Is it really fair to tax them equally? Both have the power to consume equally. But Sally leaves more for others than Connie does."

"But," asks the income tax advocate, "are you saying that if Connie consumes twice as much as Sally, she must pay exactly twice as much tax?"

"Not necessarily," replies the consumption tax advocate. "For example, under the personal consumption tax, tax rates can be set so that Connie pays more than twice the tax that Sally pays. If you want to match the distribution of the tax burden of the current progressive income tax, that's the way rates should be set. How to set the rates is a separate issue. The key point is that it is fairer to base the tax on the person's consumption rather than income."

"What about the miser with high income and low consumption?" asks the income tax advocate.

"I prefer to call him a thrifty person," replies the consumption tax advocate. "He could take a lot out of the economic pie for his own satisfaction. But he doesn't. So it is fair to charge him a low tax."

"But," continues the income tax advocate, "your thrifty person may get as much pleasure from saving as others do from consuming."

"True enough," responds the consumption tax advocate. "But we should focus on the consequences, not the motives, of each person's action. You may enjoy earning and accumulating for its own sake, but this is not sufficient reason to tax you heavily. Only when you take a huge slice of the economic pie for your own enjoyment should you be heavily taxed—not because there is anything wrong with enjoying a slice, but because it leaves less for others."

APPENDIX

A Personal Consumption Tax Return

Cash Inflows

1. Wages and salaries	$60,000
2. Interest, dividends, cash withdrawals from business	$3,000
3. Withdrawals from savings accounts or investment funds	$2,000
4. Sale of stocks and bonds	$2,000
5. Loans (excluding consumer durable loans)	$2,000
6. Cash gifts and bequests received	$1,000
7. Pension, social security, and insurance cash benefits	$0
8. Total (add lines 1–7)	$70,000

Nonconsumption Cash Outflows

9. Deposits into savings accounts or investment funds	$9,000
10. Purchase of stocks and bonds	$7,000
11. Loan repayments (excluding consumer durable loans)	$1,000
12. Cash charitable contributions and gifts given	$1,000
13. Higher education tuition (investment component)	$2,000
14. Total (add lines 9–13)	$20,000
15. *Consumption* (subtract line 14 from line 8)	$50,000

Deductions

16. Personal exemptions	$10,000
17. Family allowance	$7,000
18. Old wealth deduction	$3,000
19. Total (add lines 16–18)	$20,000
20. *Taxable Consumption* (subtract line 19 from line 15)	$30,000
21. *Tax*	$10,000
22. Payroll tax credit	$4,000
23. *Net Tax* (subtract line 22 from line 21)	$6,000

Each line of the tax return is explained in Laurence Seidman, *The USA Tax: A Progressive Consumption Tax* (Cambridge, MA: MIT Press, 1997).

IV

Policy Issues

9

The Social Contract

Once upon a time in a faraway land there lived a people in a state of nature. There were no laws to tell them what they couldn't do, no police to order them around, and no government to make them pay taxes. You might think it was heaven on earth. But, alas, it wasn't.

Each spring the Earnest family planted crops and each fall they harvested them. But each winter the Bully family ate most of the crops, while the Earnest family had barely enough crops to survive. How did this happen? Very simply. Each fall, just as the Earnests finished harvesting their crops, they were paid a visit by the Bully family.

"Here come the Bullys again!" cried the littlest Earnest, pointing out the window of their small cottage. In a few seconds the door flung open and in came the Bullys. They were big, they were mean, and there were plenty of them. "Where's the crop?" growled Big Bully himself. Trembling, the Earnests pointed to their modest barn. Soon every Bully was carrying the stored corn to the Bully wagon. In a half hour they had loaded up nearly all the crop. Suddenly the door was flung open again. It was Nasty Bully herself.

"Quit shaking and whimpering," she said with contempt. "We left you just enough to make it through the winter. After all, if you starved, who would slave for us next year?" Then she turned and cackled uproariously until she got to the huge Bully wagon. In a few seconds, the Bully wagon was gone, and so was most of the Earnest crop.

"It's not fair," cried the littlest Earnest.

"Can't we do something about it?" asked Mother Earnest.

"Not by ourselves," said Father Earnest. "The Bullys are too strong."

"But we can get help," said Mother Earnest. "We're not the only ones the Bullys pick on. They do the same thing to other families. They take tools from the Smiths, clothes from the Weavers, and a huge volume of products from the energetic Productives. They ignore the Lazys, who don't produce anything, but they take most of the meager output of the Tryers, who work long and hard but are able to produce very little."

"Let's call a secret meeting of all the families the Bullys prey on," said Father Earnest. "I think I know how we can solve our Bully problem."

Our Government

In the dead of night, while the Bullys slept soundly, the families walked silently through a moonlit field, traversed a dark woods, and finally emerged into a clearing. They lit a fire and sat near its warmth.

"I've called this meeting," began Father Earnest, "because our state of nature just isn't working. The Bullys take advantage of all of us. It's time we did something about it."

"What can we do?" whispered Timidas.

"We need to protect what we produce," replied Earnest. "I propose that each family contribute one member—a strong one—to a new organization. It will be called the *police*. Its mission will be to protect everyone from the Bullys."

"The police will need weapons," said Productive. "The Smiths make weapons. They can make them and give them to our police."

The members of the Fair family shifted uneasily. Finally Father Fair spoke up. "We don't think the whole burden should be put on the Smiths just because they make weapons. We should all share the burden."

Economas interjected, "That's easy enough. Let the Smiths make the weapons. Since every household, except the Lazys, earns coins selling what it produces, each household can contribute some coins to the police. The police will use some of the coins to buy weapons from the Smiths, and will use the rest of the coins to buy the food and clothes they need."

Suspicious spoke up. "But how do we make sure the police don't violate our rights while they track down Bullys?"

Politicas replied, "We need courts and judges to protect us from our own police, as well as to decide our disputes peacefully, and make sure only the guilty go to jail, while the innocent go free. We will need some taxes to pay the judges."

Next Mother Fair asked, "How do we make sure that everyone contributes coins, and how do we decide how much each household should contribute?"

Politicas replied, "We need to form a government. We must remember that it is *our* government, created by us, to do our bidding. It is up to us to decide what our government can and cannot do. Of course, it is impractical to have our government make decisions by calling meetings where every family participates. We're all too busy for that. Instead, we should vote to elect representatives who will meet regularly and make decisions. We know the first thing we want our representatives to do: hire and supervise the police, decide how much tax each household must pay, and make sure that every household pays its assigned tax. If they do a poor job, we'll replace them in the next election."

"Much as I hate to admit it," said Libertas, "we must give our government the power to do these things." Everyone nodded in agreement.

And so, within in a few days, the first election was held, a government formed, police chosen, and taxes collected. The next time the Bullys came to someone's house, they were shocked. Before they knew what had hit them, they were in handcuffs, riding in the new police wagon to the new jail.

Things were fine for a while, but one day the littlest Earnest climbed to the highest mountain peak in the land, looked out in the distance, squinted, and then climbed down the mountain as fast as he could and ran home shouting, "Foreign Bullys, foreign Bullys! I saw them, I saw them!"

"Tell us what you saw," said Mother Earnest, trying to calm him down.

"Far in the distance, I saw people with weapons, and horses, and big wagons. They looked just like our Bullys."

"Were they riding toward us?" asked Mother Earnest.

"Not yet," he replied. "But they could. And if they do, we'll be in big trouble."

Mother Earnest called a meeting that very night. And everyone agreed that the police could not handle an invasion by foreign Bullys. Moreover, they realized they were in great danger from foreign Bullys, because word might have spread that their own Bullys were no longer in control. Defensas said, "I propose we immediately create a *military*, far stronger than our police, to protect us against invasion by foreign Bullys."

"This will require a lot more taxes than we needed for the police," said Father Earnest.

"We have no choice," replied Defensas. "Our victory over our own Bullys will mean little if we are conquered by foreign Bullys."

And so it was agreed. Much higher taxes were raised, and a strong military was established. A message about the new military was sent to foreign lands.

A month later, the littlest Earnest climbed the same mountain and stared out in the distance in all directions. No matter how hard he looked, he saw no signs of anyone approaching with weapons, horses, and wagons. "They must have heard about our new military," he said with a proud smile.

The Free Market

Exactly one year later, the citizens met on a moonlit night in the very same clearing to celebrate the first anniversary of their new government and to assess its impact.

Productive spoke first. "Things are great. With the Bullys behind bars, my family has the skill and energy to invent and produce things people need, so this year we earned a huge volume of coins. All we want our government to do is to keep our own Bullys out of our way and prevent foreign Bullys from causing trouble."

Immediately Socialas stood up. "I couldn't disagree with you more, Productive. Now that we have formed a government, we should use it to its maximum potential. Today, every family makes its own decision about what and how much to produce. There is no central coordination and planning to make sure there is enough of good X but not too much of good Y. Every family pursues its own self-interest, trying to make as much profit as it can, without thinking about the needs of our whole society."

"What should we do?" asked Naivas.

"I propose," continued Socialas, "that the government coordinate and plan our economy. The government should draw up a detailed economic plan specifying how much of goods X, Y, and Z should be produced. Then the government should employ each family, and assign each one a specific productive task to make sure that the plan is carried out. If we don't do this, we'll have chaos."

An uneasiness came over the people. On one hand, most cherished the freedom to decide on their own economic activity without government orders. But on the other hand, Socialas's point seemed plausible. Wouldn't there be chaos unless the government specified the required amounts of goods X, Y, and Z and employed every family to make sure the required amounts were produced?

At this very moment, the most brilliant member of the Smith family, Adam, rose to his feet. In his hand he held a lengthy manuscript. "I rise to answer Socialas," Adam Smith said calmly. "But have no fear," he soothed the crowd. "I will not attempt to read this manuscript to you. I have just completed it after several years of work. It's called *The Wealth of Nations*, and I hope that some of you will find the time and interest to read it. Tonight, I will briefly summarize it in order to answer Socialas."

The people looked at Adam Smith with anticipation. He was not merely the most brilliant member of his family. Many felt he was the most brilliant person in their entire society. Then Smith began to speak.

"Each family is now free to decide what and how much to produce, and with the Bullys removed, each knows that if it produces more, it will consume more. Each is free to pursue its own economic self-interest. You all appreciate that freedom. But many of you worry that Socialas is right, that the result will be chaos. I want to assure you, this evening, that your fears are unfounded."

"By forming a government to protect us from the Bullys," Smith continued, "a remarkable system has come into being: the free market. Socialas says that chaos will result, but let me begin with this question: Has there been chaos this year?"

"Not really," said Earnest. "In fact, the economy has worked very well. All the families, except the Lazys, have been working hard to produce high-quality products at the lowest possible cost, and consumers seem pleased."

"But," asked Smith, "why do producers care about high quality and low cost? Aren't they concerned mainly about their own profit?"

"Yes," replied Earnest, "every producer strives to make a profit. I must confess that our family cares about profit too."

"Yet you say consumers seem pleased. Now, how can this be?" Smith asked with a twinkle in his eye.

"I'm not sure," replied Earnest.

"Let me ask you a question," said Smith. "How does your family decide what and how much to produce in order to make a profit?"

"Well," said Earnest, "first we think about what we're best at making. We're best at growing crops. We know we shouldn't try to make clothes or tools. The Weavers are much better at making clothes, and the Smith family—except for you, Adam—is much better at making tools."

"Yes," said Adam, "I have disappointed my family by teaching and writing instead of making tools. They are particularly upset about the subject I have chosen: economics. They say I have disgraced them by becoming an economist. But I'm hopeful that my book, *The Wealth of Nations*, will prove to be a different kind of powerful tool, worthy of a Smith. History will judge. But let's get back to your family, Earnest. How do you decide which crops will be the most profitable?"

"We see that consumers like corn better than stinkus. If we grow corn, we make a profit. If we grow stinkus, we don't."

"And why do you work hard to produce your crop at lowest possible cost?" asked Smith.

"Because we have to charge a price that covers our cost. A low cost lets us charge a low price."

"I suspect," interjected Socialas, "that you want to charge a high price even if your cost is low, so you can make a huge profit."

"I confess," said Earnest, "that you're right about what we want. But we can't do it."

"Why not?" asked Socialas. "Does your conscience stop you?"

"No, its not our conscience," replied Earnest. "It's our competition."

"I don't follow," said Socialas.

"Well," said Earnest, "we're not the only family growing corn. A dozen families grow corn. If we charge a price much above cost, they'll offer consumers a lower price, and no one will buy corn from us."

"I suspect," interjected Socialas, "that you want to grow low-quality corn because it's cheaper."

"Once again," said Earnest, "I confess that you're right about what we want, but we can't do it, not because of our conscience, but because of our competition. Most consumers are willing to pay a higher price for high-quality corn. We actually make more profit producing high-quality corn, even though the higher cost makes us charge a higher price."

"And so," said Adam Smith, "you make a profit by figuring out what consumers want, and producing it for them. Competition prevents you from over-charging and compels you to produce a high quality with the lowest possible cost."

"That's true," said Earnest.

"But," said Socialas, "he admits that his family and all the others are pursuing their own self-interest. No one is making sure that their decisions serve the public interest."

"Socialas," replied Adam Smith with a smile, "they are indeed freely pursuing their own interest, and no earthly government is telling them what or how much to produce. But in a free market they are led by an invisible hand to make decisions and pursue economic activities that serve the public interest."

The people nodded their agreement with Adam Smith, and Socialas sat down, dejected.

Social Insurance

As Socialas sat down, Father Fair stood up to speak. "Adam Smith has brilliantly shown us the virtues of a free market. Except Socialas, we are all convinced that our government should not employ families to make goods X, Y, and Z, or interfere with each family's decisions about what and how much to produce. But though we have rejected Socialas's radical role for the government, we must still ask whether there is a more limited role our government should perform as a complement to the free market."

Productive replied, "All we want our government to do is to keep our own Bullys out of our way and prevent foreign Bullys from causing trouble."

Father Fair replied, "I believe there is something else we should

instruct our government to do. It's great that the Bullys are behind bars and that the Productives can produce as much as they can without fear it will be stolen. We all benefit from their skill and energy. But not everyone has as much skill and energy as the Productives. I'm not concerned about the Lazys—at least not the grown-ups. They simply don't want to work, they earn nothing, and it's certainly fair that they consume very little, surviving on the few coins they receive from Charitas."

"But," continued Father Fair, "I *am* concerned about the Tryers. They work as long and as hard as the Productives, but they produce so much less. No matter how hard they try, they barely earn enough to survive. They need supervision, so they work for the Productives, who assign them tasks and pay them according to what they produce. You can't blame the Productives for paying them a low wage because the Tryers produce a small amount, but the low wage means they can't afford to save anything. You can't blame the Productives for laying them off when sales are slow. And when Tryers get too old to work, or need costly medical care, they're in trouble. Only the generosity of Charitas keeps the Tryers from starving when they're unemployed or too old to work, and gets them essential medical care. But Charitas is getting old, and soon there may be no one around to help the Tryers."

"That's the way it goes," replied Marketas. "The only proper role of government is to take care of the Bullys so that the free market can work. Then everyone is free to produce whatever he can and keep whatever he earns. The only thing government should do is raise enough taxes to pay the police and the military to protect everyone from Bullys and protect everyone's earnings from theft."

"But we can do better," said Father Fair. "We all agree that a state of nature is unfair, that it rewards the Bullys and harms everyone else. So we agree that we needed to create a government, and that our government should levy taxes, pay the police and military, and unleash a free market. And we agree that we should instruct our government to let the Productives earn and consume more than any other family because they produce more and we all benefit from their skill and energy. But why shouldn't we also instruct our government to help people like the Tryers?"

"The free market is natural," answered Marketas, "and the out-

comes of a free market should therefore be accepted as natural. I'm opposed to our government tampering with the natural outcomes of a free market. Such tampering is unfair and unnatural."

"But," replied Father Fair, "while the free market is wonderfully productive, it is not natural. In the state of nature there was no free market. The Productives suffered along with everyone else. It was the Bullys who triumphed in the state of nature. We needed to invent our government in order to create a free market. But once we recognize that the free market is not natural and depends on our government, it makes sense to ask whether we should instruct our government to do anything else. Shouldn't we use our government to help the Tryers as well as helping the Productives?"

Marketas shook his head. "The only way to make the Tryers better off is to make the Productives worse off. You could instruct our government to take so many coins from the Productives and give so many coins to the Tryers that they would consume equally. But not only is that unfair—after all, we all benefit more from the Productives than the Tryers—but it would destroy the incentive of people to become as productive as they can be. Why make the effort to develop skill if you will not be allowed to consume the fruits of your higher skill? And if potentially productive people don't develop their skill, we will all miss the products they could have invented but didn't."

"I agree, Marketas," replied Father Fair. "If we instructed our government to tax the Productives so much and give so much to the Tryers that they both consumed equally, the outcome would be both unfair and disastrous. Neither the Productives nor the Tryers would work at all. But I'm proposing something much less extreme."

"Tell us," said Earnest.

"We should instruct our government," replied Father Fair, "to set up three social insurance programs: unemployment insurance, Social Security, and health insurance. The first would pay benefits to the unemployed; the second, to the old; and the third, to the sick. Our government would raise the coins it needs for social insurance by taxation."

"You better not be too generous with these social insurance benefits," warned Incentivas. "If you give an unemployed person a benefit as high as her former wage, she won't try to get another job. If you give an old person a benefit as high as her former wage, she

won't save anything when she's working age, and our banks won't have the funds to lend to businesses to make investments in technology from which we all benefit. And if you pay the entire medical bill of a sick person, she will have no incentive to consider cost when she orders medical care and we will all pay inflated taxes to cover her inflated medical costs."

"Your warnings are wise, Incentivas," replied Father Fair. "We must strike a balance. The unemployed person must get a benefit that is only a percentage of her former wage, so she can survive but still has a strong incentive to find another job as quickly as possible. The old person must get a benefit that is only a percentage of her former wage, so she can survive but still has a strong incentive to do some saving when she is working age. And the sick person must get a benefit that is only a percentage of her medical bill, so she can afford medical care she needs but still has an incentive to weigh the cost of anything she orders."

"Would these percentages be the same for everyone?" asked Earnest.

"I think it would be better," replied Father Fair, "to vary the percentages. Low earners like the Tryers should be given a high percentage, because despite their effort, their wage is low and they can hardly afford to save anything. Low earners like the Tryers might receive a benefit for unemployment or old age that is 60 percent of their former low wage, and a benefit that is 95 percent of their medical bill; high earners like the Productives might receive a benefit for unemployment or old age that is 20 percent of their former high wage, and a benefit that is 75 percent of their medical bill. In other words, I favor a *progressive* benefit schedule for social insurance programs—the lower the earnings, the higher the percentage."

"I have one problem with your medical percentage," said Earnest. "Some unfortunate people incur enormous medical bills. The low earner may not be able to afford to pay even 5 percent of the whole bill, and even the high earner may be unable to afford paying 25 percent of the whole bill."

"You're right," replied Father Fair. "Under my medical benefit schedule, once a household has paid an amount equal to a certain percentage of its income, it would not have to pay any more that year."

"Would this percentage of income be the same for everyone?" asked Earnest.

"Once again," said Father Fair, "I think it would be better to vary the percentage. For example, it might be 1 percent of income for the Tryers, but 5 percent of income for the Productives."

"But why should we instruct our government to provide insurance?" asked Marketas. "Insuras and her family do a nice job of selling private insurance under our free market. Why can't we rely on private insurance for unemployment, old age, and medical care?"

Economas turned and faced Marketas. "My dear Marketas, you know I teach the virtues of relying on the free market for almost everything we make. I argue vigorously against Socialas who wants our government to employ everyone and make everything. We need a free market with competition, not a government monopoly, to have a productive economy. But our beloved free market does a poor job with certain kinds of insurance. Tell me, what kind of insurance does Insuras sell?"

"Mainly wagon and home insurance," answered Marketas. "With wagon insurance, you're covered if your wagon injures someone and you owe huge damages. With home insurance, you're covered if a fire destroys your home."

"Yes," said Economas, "Insuras does a nice job with wagon and home insurance. But does Insuras sell unemployment insurance?"

"Come to think of it," answered Marketas, "she doesn't."

"No, she doesn't," said Economas. "And do you know why not?"

"No," answered Marketas, puzzled.

"Because," continued Economas, "she would be glad to sell unemployment insurance to the Productives even at a low price, because they'll probably never become unemployed. But the Productives won't buy insurance, because they can handle a brief unemployment spell with their own saving. On the other hand, Insuras won't sell insurance to the Lazys, who would immediately start collecting forever. She might like to sell to people like the Tryers, but she can't always tell whether a person will act like a Tryer or a Lazy. If a person wants to buy insurance, there's a good chance he'll act like a Lazy. So Insuras would have to set a high price. But then the Tryers won't think it's a good deal, only the Lazys will buy it, and Insuras will have to raise the price even further. Eventually, even the Lazys will think it's a bad deal. That's why Insuras doesn't sell unemployment insurance. We need our government to provide it."

"But," replied Marketas, "Insuras does sell old age insurance (which requires contributions while the person works and during retirement pays benefits for as long as the person lives), and medical insurance."

"Unfortunately, Marketas, there are several problems," replied Economas. "Let me begin with old age insurance. People like the Tryers simply don't earn a high enough wage to be able to afford to save enough for retirement. They need most of what the Productives pay them to subsist while they're working. So they can't afford to save enough or to buy old age insurance from Insuras.

"Then there are families like the Unluckies," continued Economas. "They save while they work, but misfortune strikes. A few years before retirement they happen to lose their jobs through no fault of their own, and they're forced to use up their savings until they find another job. Or they put their savings into stocks of companies that unexpectedly have difficulty selling their products, so the value of their stock plummets.

"Finally, there are the Myopics, who earn a high enough wage, but they just don't think about what will happen when they get old. They spend all their coins today, and when they get old, they find themselves in trouble. Perhaps they deserve to starve in old age, but many of us (especially Charitas) won't let them—after all, they do work hard up until retirement. So we end up bailing them out with our coins, and they get away without saving."

Mother Fair interjected, "Rather than the rest of us bearing their burden, it would be fairer to make the Myopics contribute coins while they're working rather than letting them get a free ride. Only our government can make them bear their fair share of the burden by taxing them while they work."

"So," concluded Economas, "while some families save successfully on their own or buy enough old age insurance from Insuras, many do not. It makes sense to instruct our government to provide old age insurance so that everyone who worked when they were younger lives decently in retirement. Everyone would earn old age protection by paying taxes during their work life. If a family wants to live better than that, it must save on its own."

"Would the taxes of workers go straight out to retirees, or instead be invested in stocks and bonds to be used when workers retire?" asked Incentivas.

"That's a good question," replied Economas. "We should discuss the pros and cons of each method at a future meeting. Here I want to argue only that our government should provide old age insurance financed by our taxes."

"What about medical insurance?" asked Marketas.

"Yes," replied Economas, "some families buy medical insurance from Insuras. But, again, the Tryers can't afford it. And some families, like the Unhealthies, are not offered insurance at an affordable price. One of the Unhealthies has a chronic costly medical problem. You can't blame Insuras for refusing to offer insurance to the Unhealthies, because Insuras would lose a lot of money enrolling such costly families."

"Any family," said Mother Fair, "could suddenly come down with a chronic, costly medical problem. Would that family lose its insurance?"

"Not immediately," replied Economas. "But Insuras purposely limits the insurance contract to one year. You can't blame Insuras for refusing to renew the insurance as soon as the year is up."

"True, I can't blame Insuras," said Mother Fair, "but I find this situation very disturbing. Any family can lose insurance just when it needs it most."

"But," interjected Marketas, "the situation is not as bad as that. Insuras often sells insurance to an employer, and agrees to cover and automatically renew all employees, whatever their medical costs. It's worth it to Insuras because it is profitable to enroll a large number of families all at once."

"You are quite right, Marketas," continued Economas. "As long as you can keep your job in a large business, your family is safe. But if you leave that workplace because you're too sick or too old, or become self-employed, employed by a small business, or unemployed, you're in trouble. Insuras will either charge you a very high price or refuse to sell you insurance."

"That could happen to any of us," said Mother Fair. "It will definitely happen once we get too old to work."

"So," concluded Economas, "while some families successfully obtain renewable medical insurance as long as they work for a large business, many do not. And even these families worry that they may lose this employment and medical insurance. It makes sense to in-

struct our government to provide medical insurance so that every family, lucky or unlucky, knows it will always be able to obtain needed medical care. Remember, every patient would be required to pay a percentage of her own medical bill until the burden becomes too great in a single year. Our government should provide universal, fair medical insurance, not free medical care."

The Social Insurance State

"What about the Lazys?" asked Earnest. "They're able to work but they just don't."

"I think we should help them too," said Compassionas. "They have a need for food, clothing, and shelter, just like the rest of us. My motto is, 'To each according to his need.' So I think we need a fourth program, 'welfare,' to give coins to the Lazys."

Mother Fair responded, "True, the Lazys have a need, but they also have a responsibility. If we all refused to work, we would all starve, including the Lazys. It would be unfair to give coins to the Lazys when the Tryers work so hard to earn enough coins to survive. Welfare should give coins only to people genuinely unable to work due to physical or mental incapacity."

"I agree about the Lazys," replied Earnest. "But there are some people who want to work, but they have low skill, and have trouble finding an employer who finds it profitable to hire them even at a subsistence wage."

"The Lazys say they want to work, but they just can't find a job," said Naivas.

"There is only one way to put someone to the test," said Economas. "Our government should provide last-resort low-wage jobs, not welfare, for anyone physically and mentally able to work. If someone can't find a job with a regular employer, that person should be given a last-resort low-wage job."

"But what jobs can the government offer these low-skilled people?" asked Earnest.

"There's plenty of low-skilled work that needs to be done," replied Economas. "Our parks and streets need constant cleaning. The walls of our buildings need graffiti removed."

"But can the government efficiently operate a jobs program?" asked Earnest.

"It might be better," said Economas, "to have the government contract with private firms to run the work projects. But the government must pay for the projects and make sure that enough are available so that anyone who wants to work gets the chance."

"Why must the jobs have a low wage?" asked Naivas.

"Because this work doesn't deserve a wage as high as the Tryers earn from a private employer," said Mother Fair.

"Moreover," added Economas, "we want everyone to have an incentive to prefer regular jobs to these last-resort jobs."

"But how can they survive on such a low wage?" asked Naivas. "Even the Tryers can barely survive."

"Our government should give a supplement to all low-wage workers," answered Economas, "whether the person works for a regular employer or in the last-resort jobs program. The supplement wouldn't be welfare because no one would get it unless they worked and earned a low wage; up to a point, the more they earned, the more supplement they would get. We can efficiently administer this supplement through our tax system, so I propose that we call the supplement the Earned Income Tax Credit. I'll explain the details at a future meeting."

"So," Earnest said, "our message will be this: There's no welfare if you can work but don't. But there's always a place to earn enough to survive: the last-resort jobs program. Though the wage is low, if you work hard, your supervisor will give you a letter of recommendation to help you get a higher-paying job with a regular employer."

"But what if someone gets the last-resort job," said Tryer, "but then makes little effort to do the job right?"

"Then that person must be fired," said Mother Fair, "and be given another chance a month or two later. A month or two without coins might change the attitudes of those who shirk the first time."

"But what about their children?" objected Compassionas. "Should children suffer because their parents won't work?"

"You ask a tough question," replied Mother Fair. "Remember, the children will be covered by our medical insurance program, because that will be universal and automatic. But they will still suffer if their parents refuse to work."

"We can't let lazy people extort coins from us by threatening to let their children starve," said Tryer.

"I agree," said Mother Fair. "Perhaps the best solution is to treat repeated shirkers as being guilty of child neglect. We already remove children from parents who severely neglect or abuse them. Refusing to take a job that can give your child food is surely severe neglect. Of course, this must be handled carefully, with due process and sensitivity. Clear warnings must be given. No child should be taken from a parent unless the child is in genuine danger. And a parent must be able to get the child back by working responsibly."

"So our social insurance programs would not make us a 'welfare state,'" said Earnest. "There would be welfare only for those physically or mentally unable to work. Everyone else would have to work to earn coins. Perhaps we should call ours a *social insurance state.*"

"I agree," said Mother Fair.

Educas spoke up. "I very much support our social insurance programs and jobs program. But we need one thing more. Education helps make an economy productive, but just as important, it is the key to a child's opportunity. Up until now, our schools have been private, and parents have had to pay the entire tuition themselves. But some families earn enough coins to pay the tuition of a high-quality school, while others earn barely enough to pay the low tuition of a low-quality school."

"This is wrong," said Mother Fair. "We should reduce the quality gap among our schools."

"The solution," said Educas, "is to levy taxes to pay for schools. The tax each family is assigned should vary with the family's income. Tax finance will enable children of all families to attend schools with decent quality."

"This doesn't mean," noted Economas, "that the government must use all tax revenue to finance free public schools. Some of the revenue should be used for scholarships for nonaffluent families, so they have the ability to switch to a private school if they are very dissatisfied with their public school. That's a subject for a future meeting."

A Progressive Tax

"I propose," said Father Fair, "that the government use a *progressive* tax to raise the coins to pay for the military, the police, the judges, the schools, and the last-resort jobs program. It's all right for the

social insurance programs to be financed mainly by an earmarked wage (payroll) tax so that each working person knows he has contributed and has earned insurance benefits through work. But for all other government expenditures, we should use a progressive tax."

"What's a progressive tax?" asked Earnest.

"It's a tax that takes a higher percentage from the affluent than from the nonaffluent," answered Father Fair. "Economas, would you give us an example?"

"Certainly," replied Economas. "To keep it simple, let's assume there is only one Productive and one Tryer. Productive earns 100,000 coins this year, while Tryer, working just as long and hard, manages to earn only 10,000 coins, so the ratio of before-tax incomes is 10 (100,000/10,000). A proportional tax would take the same percentage from everyone; for example, a 20 percent proportional tax would raise 20,000 coins from Productive and 2,000 from Tryer, for a total tax revenue of 22,000. Productive would pay 10 times as much tax as Tryer, and Productive would have 80,000 coins left, while Tryer would have 8,000 left. Note that with a proportional tax, the ratio of after-tax incomes is 10 (80,000/8,000), the same as the ratio of before-tax incomes."

"How would the progressive tax be different?" asked Naivas.

"Under a progressive tax," said Economas, "Productive would pay a higher percentage than Tryer. For example, instead of both paying 20 percent, suppose Tryer pays 12 percent while Productive pays 20.8 percent. Then Tryer would pay 1,200 coins (.12 × 10,000) and Productive would pay 20,800 coins (.208 × 100,000), again for a total tax revenue of 22,000."

"Is that fair to Productive?" asked Arithmetas. "I just calculated that Productive would pay 17.3 times as much tax as Tryer (20,800/1,200 = 17.3), even though Productive earns 10 times as much income as Tryer."

"But," said Economas, "can you tell us, Arithmetas, how many times more would Productive be able to consume than Tryer?"

"That's easy enough," answered Arithmetas. "Productive will have 79,200 coins left after tax, while Tryer will have 8,800 coins left, so Productive will be able to consume 9 times as much as Tryer (79,200/8,800 = 9). That's remarkable. Even though the progressive tax makes Productive pay 17.3 times as much tax as Tryer, it still lets Produc-

tive consume 9 times as much as Tryer. That's not much less than the earnings ratio of 10."

"But," warned Marketas, "that drop from 10 to 9 may weaken the incentive of Productive to work his hardest."

"I would still work my hardest if I could consume 9 times as much," said Naivas.

"I admit," confessed Productive, "that I would still work my hardest. But you're missing the point. I earned 100,000 coins in the free market. Government has no right to take coins from me. The coins are my property. I'm entitled to consume 10 times as much as Tryer because I produced and earned 10 times as much."

"Aren't you overlooking one thing, Productive?" said Father Fair. "Without a government, taxes, and police, you would consume the same amount as Tryer, thanks to the Bullys—barely enough to avoid starvation. So if the government enacts a progressive tax where you pay 20.8 percent and Tryer 12 percent, you would consume 9 times more than Tryer. While 9 times under the progressive tax is not quite as good as 10 times under the proportional tax, it is still much better than the state of nature, where your consumption would be the same as his."

"I'd take 9 times," said Naivas.

"But it's still not fair," insisted Productive. "I produce 10 times as much, not 9 times as much."

"Yes," replied Father Fair, "you are blessed with the potential to develop high skill, and you work hard to develop it. We all benefit more from you than we do from Tryer, so I agree that you deserve to consume more than Tryer. But 9 times is certainly a lot more, even if it is less than 10."

"I think the Tryers deserve to consume as much as the Productives," said Egalitas. "Without government, they would consume the same, thanks to the Bullys. The Tryers work as long and as hard as the Productives. It's not their fault that they weren't given the potential to develop high skill. I say the government should tax the Productives more, and transfer coins to the Tryers rather than tax them, so that each Tryer ends up with the same coins as each Productive."

"Egalitas, I disagree with you for two reasons," replied Father Fair. "First, I don't think effort is all that matters for fairness. True, the Tryers give the same effort as the Productives. But the fact is that the

rest of us benefit more from the Productives than we do from the Tryers. So I think it is fair to weigh actual productivity as well as effort. Second, your proposal is impractical. It would destroy the incentive of everyone to work hard, and we would all lose. Each Productive will work just as hard if each gets to consume 9 times, instead of 10 times, as much as each Tryer. But if we make the tax-transfer system much too progressive—if we go to your extreme of making everyone end up with equal consumption—then we will end up with poverty for everyone because no one will work hard."

A Progressive Income Tax or a Progressive Consumption Tax?

"I'm persuaded that our tax should be progressive," said Earnest. "But that still leaves open this question: Should the progressive rates apply to each family's income or to its consumption?"

"Good question," said Economas. "Which one we choose could have an important effect on our future standard of living. If families are taxed on their consumption, not their income, there will almost surely be more saving in our economy. This saving is put into banks that lend it to our business firms so they can invest in new machinery and technology. Saving is necessary to finance investment, and investment is what makes our productivity and standard of living rise."

"But is it fair to tax each family according to its consumption, rather than its income, even at progressive rates?" asked Mother Fair. "After all, consider the thrifty person who earns a high income but consumes little. He has a high ability to pay. Shouldn't he pay tax according to his ability?"

"Some are surely persuaded by this ability-to-pay argument," replied Father Fair. "But there is another way of looking at it. When a thrifty person takes only a small amount out of the economic pie to consume for his own enjoyment, he leaves more resources for others to consume, or for businesses to invest. That investment raises the future productivity, wages, and standard of living of everyone. True, he may get pleasure from saving, while others get pleasure from consuming. But the consequence for the rest of us is very different. The more someone saves, the higher is everyone else's future standard of living."

"So," said Mother Fair, "you believe that consequences are relevant to fairness, and therefore it is fairer to tax families according to what they take out of the pie for their own consumption."

"Yes," replied Father Fair.

"In that case," said Earnest, "it doesn't make sense to tax people when they transfer wealth through gifts or bequests, or when they inherit wealth."

"I agree," replied Father Fair. "Giving or receiving wealth does not remove resources from the economic pie. It does not reduce the resources available for businesses to invest. Only when someone consumes his wealth should he be taxed, and then at progressive rates."

"Not having wealth transfer taxes would surely be simpler," said Earnest. "But is it practical for each household to compute its annual consumption?"

"It is," replied Economas. "Each year a household must add its cash inflows, like wages and withdrawals from bank accounts, and subtract nonconsumption cash outflows, such as deposits in saving or investment accounts, or purchases of stocks and bonds. What's left is the household's consumption. Of course, it's more complicated than this. Some things are more complicated under a consumption tax, but others are more complicated under an income tax. I propose we devote a future meeting to these practical aspects. What is certain is that our future standard of living will be higher if we use a progressive consumption tax instead of a progressive income tax."

"And though some disagree," said Father Fair, "in my view our tax will be fairer if we choose a progressive consumption tax over a progressive income tax."

Our Constitution

At the next annual meeting to review the progress of their government, Politicas spoke first. "We should be proud of ourselves. We saw that the state of nature wasn't working, so we came together to form our government. We immediately agreed on the need for taxes to pay our police and judges, and soon after agreed on even higher taxes to pay our military. Socialas made the radical proposal that our government employ everyone and run our economy, but Adam Smith showed brilliantly why we should let the free market guide our

economy instead. After Smith spoke, no one except Socialas wanted our government to employ everyone to make all goods and services. Some of us wanted to limit our government to the police, courts, and military. But the majority wanted our government to use taxes to finance several social insurance programs, a last-resort jobs program, welfare only for people genuinely unable to work, and elementary and secondary education. The majority opted for a progressive consumption tax to finance all government expenditures except the social insurance programs, which would be financed by a payroll tax.

"Instead of remaining in a state of nature," continued Politicas, "we entered into a social contract, specifying what our government would and would not do. Now that we have experimented, I propose that we embody our social contract in a written constitution. This will prevent disputes in the future about what our government can and cannot do, and ensure our rights and liberties. We must be humble enough to allow a procedure for amending the constitution. But amending it should be hard, not easy, so that it will occur only when a large majority concur."

Politicas was nominated to head a committee to draft a written constitution embodying the social contract. On the day the constitution was ratified by the people, the state of nature was declared gone forever, permanently replaced by a social contract utilizing a government created by the people for their benefit and guided by a written constitution. It was a day of celebration.

10

Social Security

Will today's twenty-year-olds get any Social Security benefits when they retire after 2040? According to polls, many twenty-year-olds think they won't. They're wrong. Below I'll explain why they'll get substantial Social Security benefits. But does this mean today's twenty-year-olds shouldn't worry about Social Security? No, they should worry. Unless something is done, after 2040 the Social Security "replacement rate" will be 15 percent lower than today; instead of receiving a benefit that is 40 percent of preretirement wage income, the benefit will be only 34 percent (6/40 = 15%). Moreover, payroll tax rates on future workers will have to be about 20 percent higher than today just to achieve this 34 percent replacement rate; the combined employee plus employer rate will have to be 15.0 percent instead of today's 12.4 percent (2.6/12.4 = 21%). So future retirees and workers after 2040 will be unhappy with Social Security because the replacement rate will be 15 percent lower, and the payroll tax rate, 20 percent higher, than on previous generations.

Can anything be done today to reduce this financial crunch after 2040? To see what should be done today for Social Security, think about what you should you do if you foresee a financial crunch in your future when you retire. The answer is obvious though unpleasant: first, you should plan to work to an older age and retire later; and second, you should save more while you work, gradually building up

a big fund that you can draw on when you retire. Of course, working to an older age and saving more while you work are unpleasant. So it's understandable that you would try desperately to find some other solution. But there isn't any. The same is true with Social Security.

There are two ways to save more for retirement. The first way is to build up one large Social Security fund; the second way is to have each individual build up her own individual Social Security fund on top of regular Social Security. Both ways will work, but both involve sacrifice. Below, I will discuss the differences between the two approaches. But first some background.

Background

Social Security was enacted in the United States in 1935 as the crowning jewel of President Franklin D. Roosevelt's "New Deal." The original plan was for workers (and their employers) to pay payroll taxes (a percentage of wage income up to a ceiling) and have one large Social Security fund build up, and then when these workers retired they would receive benefits from the single large fund they had helped to build. So in the late 1930s, payroll taxes were paid and the single fund began to build up, but few benefits were paid out because few retirees had paid into the fund while they were working—according to the original plan, they were therefore not eligible for benefits.

But there was a lot of sympathy for the retirees of the late 1930s. The Great Depression had forced many of them to use up their savings when they became unemployed. So many, through no fault of their own, faced old age without savings. Couldn't anything be done to help them? The answer was yes. Instead of sticking to the original plan, Social Security could take the payroll tax revenue coming in and use it to send benefits out to these unfortunate retirees. But immediately the question arose: What will become of today's workers when they retire, because there won't be any build up of a large fund? And immediately came an answer: When they retire, the payroll taxes of the next generation of workers (and their employers) will be used to pay their benefits. And so by 1940 Social Security adopted the new plan, and began paying out benefits instead of building up a large fund. The new plan came to be called "pay-as-you-go (PAYGO)" because when people worried how benefits would be paid if no fund

were built up, they were told, "Don't worry, we'll pay as we go—we'll use tomorrow's payroll taxes to pay tomorrow's benefits."

Now, needless to say, the new plan was a great deal for people retiring in the 1940s. Take Ida Fuller of Vermont. Legend has it that she was about to retire when Social Security was enacted, so prior to retirement she (and her employer) paid only about $20 of payroll tax. But then she had the fortitude to live to a very old age, collecting about $20,000 in Social Security benefits! Before getting too angry at Ida and other retirees of that era, remember that most of them suffered economic hardship through no fault of their own due to the Great Depression. At any rate, even if you are still angry, there's nothing you can do about it. These retirees have long been dead, so there is no way to get the money back from them.

For the next forty years until 1980, Social Security followed the PAYGO plan. Each year, payroll tax revenue would be sent out to pay benefits. For people retiring in the 1950s, Social Security was not quite as spectacular a deal as it had been for people retiring in the 1940s, but it was still a very good deal: the 1950s retirees had only paid payroll tax during the last decade or two of their work careers, but were now receiving benefits financed by payroll taxes on the entire work force. Each decade, Social Security became less of a great deal. By 1980, virtually every retiree had paid payroll taxes over his entire work career, so for each retiree, the taxes he had paid were no longer much less than the benefits he received during retirement.

Not only was Social Security no longer a spectacular deal, but a demographic threat loomed on the horizon. Around 2010, the numerous baby boomers—born in the decade and a half following World War II—would start switching from being workers who paid payroll taxes to retirees entitled to Social Security benefits. Moreover, medical advances would enable many of these boomers to live to a very old age—they would spend many years in retirement receiving benefits.

What could be done to reduce the severity of this coming financial crunch for Social Security? In a bipartisan manner, Congress and the president rose to the challenge. In the early 1980s, a bipartisan commission, headed by Alan Greenspan (who did his job so well he was appointed chairman of the Federal Reserve in 1987), recommended

raising payroll taxes to build up the Social Security Trust Fund. Rather than hold cash, the fund would sensibly hold safe U.S. Treasury bonds that would pay interest, and the interest income would help pay benefits when the baby boomers retired. Cynics said the fund couldn't be built—that Congress would never vote for a payroll tax increase to build it, and even if it did, that Congress would soon dissipate the fund. But the cynics were wrong. With bipartisan support—so that neither political party could blame the other—Congress passed the payroll tax increase—and over the next two decades, the fund grew steadily so that there are now over a trillion dollars worth of U.S. Treasury bonds held by the Social Security Trust Fund, earning substantial interest income every year. In 2003, the combined employer plus employee payroll tax rate was 12.4 percent applied to an employee's wage income up to an annual ceiling of $87,000 (this number is automatically raised each year to keep pace with increases in wage income).

Unfortunately, the build up, though impressive, is still not large enough. The Social Security actuaries report that if the current tax and benefit rules are unchanged, then in about two decades the trust fund will have to start gradually cashing in its Treasury bonds to get cash to pay promised benefits—tax revenues and interest won't be enough to pay promised benefits—and by 2040 there will be no bonds left in the trust fund. This is shown in Figure 10.1. So what will happen after 2040? Will Social Security stop paying benefits? Of course not. At current tax rates, every year Social Security will continue to collect a huge amount of payroll tax revenue that it will immediately send out to retirees to pay benefits. Anyone who thinks there will be no Social Security benefits when the trust fund runs out of bonds after 2040 doesn't understand the simple fact that payroll taxes collected each year, not bonds or cash piled up in the trust fund, are the main source of revenue to pay benefits.

But there will be a very serious problem after 2040. Social Security actuaries estimate that, with current tax rates, the payroll tax revenue collected after 2040 will be only about 70 percent of promised benefits under current benefit rules. That's why, prior to 2040, the trust fund will have to cash in some of its Treasury bonds every year to be able to pay promised benefits. So when the bonds are gone, there will be a 30 percent gap. What will happen? At one extreme, Congress could raise $100 instead of $70 in payroll taxes and keep

Fugure 10.1   **Social Security Trust Fund**

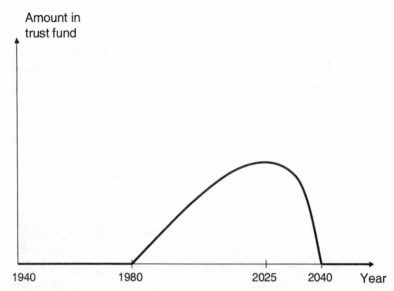

Amount in trust fund

1940   1980   2025   2040   Year

benefits unchanged; at the other extreme, Congress could cut benefits from $100 to $70, keeping payroll taxes unchanged. Most likely, Congress will split the difference: It will probably raise payroll taxes from $70 to $85 (a 20 percent increase, from the current combined employer plus employee rate of 12.4 percent to 15.0 percent), and cut benefits from $100 to $85 (a 15 percent cut, from a current replacement rate of 40 percent to 34 percent).

So after 2040, both retirees and workers will be very unhappy with their deal under Social Security. There will be an unpleasant intergenerational tug-of-war fought out in Congress, with lobbying by the old against benefit cuts and lobbying by the young against payroll tax hikes. Congress will compromise, and neither side will be happy. There will be no collapse, but there will be much bad feeling, and much dissatisfaction with Social Security.

Two questions will naturally arise after 2040: Why do we retirees deserve a replacement rate much smaller than previous retirees? And why do we workers deserve payroll tax rates much higher than previous workers? Good questions. Why should Social Security treat the retirees and workers after 2040 much differently from the retirees and workers before 2040?

Demographics, Longevity, and the Rise of the "Big Three"

For many years, federal government spending and taxes have each been about 20 percent of GDP. In 2000, over a third of federal spending, 8 percent of GDP, consisted of the "Big Three": Social Security, Medicare, and Medicaid (hence, the rest of federal spending was 12 percent of GDP). Medicare is medical spending for retirees. Medicaid is medical spending for low-income people, many of whom are retirees. In 2000, the Congressional Budget Office issued a study in which it projected that, because of the retirement of the numerous baby boomers, and the fact that boomers will be living to older ages, federal spending on the Big Three (under current benefit formulas and the current retirement age) will rise from 8 percent of GDP in 2000, to 12 percent in 2020, to 16 percent in 2040. If the rest of federal spending remains 12 percent of GDP, federal spending will rise to 28 percent of GDP, way above its historical 20 percent. As a consequence, either federal taxes will have to rise to a historic height of 28 percent of GDP, or the federal government will have to borrow huge percentages of GDP each year, going deeply in debt and incurring huge interest payments every year.

What should be done? To see the answer, get personal. Suppose you suddenly discover that you are going to live to a much older age. What should you do? Two things: first, raise your retirement age, and second, save more while you're working. The answer is the same for Social Security. Let's consider each in turn.

Today, most people live longer and work jobs that are less physically demanding than years ago. Hence, it makes sense to gradually raise the Social Security retirement age above 65. In fact, Congress has already done this. In the early 1980s, the same bipartisan commission mentioned above recommended raising the retirement age to 67, and Congress enacted a schedule for a very gradual phase in. Under that schedule, beginning in 2000 the retirement age rises 2 months per year until it reaches 66 in 2005; the phase in is then suspended for ten years; and then the 2 month per year phase in resumes so that the retirement age reaches 67 in 2022. So what else can be done? The answer is obvious: Skip the ten-year suspension so that the retirement age of 67 is reached ten years earlier in 2012. It would

then be possible to continue the 2 months per year phase in after 2012 so that, for example, the retirement age reaches 68 in 2018. Note that a person could still be given the option of retiring as early as 62 and beginning Social Security at that age (just as a person can today), but the monthly benefit would naturally be reduced for each month that the person starts before the regular retirement age.

Second, you must save more while working. To make sure you save, the government can raise your Social Security tax. Then there are two options. Either the government can save for you by using your tax to build up the Social Security Trust Fund. Or the government can let you save in your own individual Social Security account.

There are two ways to raise payroll taxes for Social Security. First, the payroll tax rate—the percentage—can be raised. Today the combined employer plus employee rate (excluding Medicare) is 12.4 percent. That rate could, for example, be raised 1.5 percentage points to 13.9 percent. Second, the payroll tax ceiling can be raised. In 2003, the payroll tax ceiling was $87,000—this means that no more than $87,000 of a person's annual wage income was subject to the 12.4 percent tax. The ceiling could, for example, be raised $10,000 to $97,000. In the early 1980s, Congress used both methods to raise payroll tax revenue. The first method—raising the tax rate—puts some burden on all workers. The second method—raising the ceiling—puts the burden only on workers earning more than the current ceiling.

One Large Fund Versus Many Individual Funds

But should your tax go to the Social Security Trust Fund, or to your own new individual Social Security fund? Of course, building many individual funds on top of regular Social Security is just as painful as building up the single Social Security fund. Both require you to pay more tax.

Once we recognize that only an approach that involves sacrifice will work, we are ready to ask: "So which is better: a single large fund, or many individual funds?" I will simply delineate the differences so that you can decide which you prefer.

With individual funds, each person would choose how to invest

her own account and take the consequences. One person might play it safe with government bonds. Another might take a chance on corporate stock. Each individual, not the government, would make the decision. If an individual invests in corporate stock, she is on her own; she may achieve a high return or suffer a low return. With a single fund, investment risk is pooled for all retirees. If the single fund plays it safe by investing solely in government bonds—as it is required to do under current law—there is little risk. If Congress changes the law so that the Social Security Trust Fund was permitted to invest a portion in corporate bonds or even stocks, there would be some risk, but the risk would be pooled over all retirees.

Establishing and managing new individual funds would involve new administrative costs because new services must be provided to each individual. Each person must be able to periodically change his portfolio of stocks and bonds, regularly receive reports of his fund's earnings, and upon retirement either choose to purchase an annuity that will pay regular benefits as long as the person lives, or select a schedule for drawing down his fund (leaving a bequest to heirs if he dies before the fund is exhausted).

With a single fund, workers with the same wage histories would receive the same benefits because benefits are based on each person's wage history according to the benefit formula enacted by Congress. By contrast, with many individual funds, workers with the same wage histories would receive different benefits because returns vary among individual funds. If one person chooses corporate stocks that rise in value, and another, stocks that fall, their benefits in retirement would differ correspondingly. If Social Security established new individual accounts, a policy decision would have to made about whether and how to guarantee each person a minimum benefit, or to limit the percentage of his fund that a person could invest in risky stocks or bonds.

A single fund would permit some redistribution whereas individual accounts would not. Under Social Security's progressive benefit formula, there is some redistribution from high-wage to low-wage workers when they retire: If person H earned three times the wage and paid three times the tax each year as person L, person H would receive a benefit perhaps twice but not three times as great as person L because of the progressive benefit formula. By contrast, with indi-

vidual funds, if person H earned three times the wage as person L, person H would receive a benefit approximately three times as great as person L.

An argument against a single fund is that the some people are skeptical that the bonds owned by the Social Security Trust Fund will be used for Social Security rather than for something else, despite reassurance by most experts that the bonds are safe. Currently, the Social Security Administration, located in Baltimore, does not actually hold marketable government bonds in its vault. Instead, it trusts the U.S. Treasury in Washington D.C. to hold its bonds for it, and the bonds in D.C. are special nonmarketable bonds. A simple, clever way to strengthen the public's confidence would be for Congress to instruct the U.S. Treasury to replace the special bonds with regular marketable U.S. government bonds, and deliver these bonds to the Social Security Administration in Baltimore to hold in its own vault. The Social Security Administration could then periodically report to the public the value of the bonds it has in its vault.

Another argument against a single fund is that government bonds won't earn as high a return as corporate stocks. This argument was popular in the late 1990s as the stock market soared, but the plunge in the stock market in the early 2000s reminded everyone that stocks can be risky. Perhaps it's best after all for the Social Security Trust Fund to hold only safe U.S. government bonds. But if Congress ever decides that the trust fund should be permitted to hold some corporate stocks and bonds, it can instruct the Social Security Administration to contract with private investment firms to invest a portion of the trust fund in a conservative diversified portfolio of stocks and bonds.

To summarize: As long as you pay more tax—for example, an amount equal to 13.9 percent of payroll instead of 12.4 percent— you will help prepare for the future, and reduce the 2040 problem, whether your tax goes to build up the Social Security Trust Fund, or to build up your own individual Social Security fund.

But individual accounts won't work if they are a "carve out" instead of an "add on." If the 1.5 percentage points for your new individual account is "carved out" of the current 12.4 percent, then only 10.9 percent will be left for regular Social Security. In the future, the benefit from regular Social Security would be lower (and the trust

fund would be exhausted earlier) so a carve out would make little difference for the 2040 problem. So if you're asked whether you favor new individual accounts under Social Security, be sure to reply: "As an add on, or as a carve out?" An add on involves sacrifice, and will therefore work to mitigate the 2040 problem. A carve out involves no sacrifice, and will not work.

Conclusion

Social Security faces a serious problem after 2040. Unless something is done today, the replacement rate may have to be cut from 40 percent to 34 percent (a 15 percent cut) and payroll taxes raised from 12.4 percent to 15.0 percent (a 20 percent increase). The only way to reduce the 2040 problem is to sacrifice: first, work to an older age, and second, save more while working. There are two ways to save more: through a single fund or many individual funds. Both ways will work, *provided* the individual funds are an "add on" on top of the current 12.4 percent (for example 1.5 percentage points for a total of 13.9 percent), not a "carve out" from current 12.4 percent (that leaves, for example, only 10.9 percent for regular Social Security and keeps the total at 12.4 percent). Both ways will work, but they have important differences. Each citizen should check the differences, and decide which solution she prefers. But there is no escaping sacrifice if we are to reduce the 2040 problem.

11

Health Insurance

As his plane landed in Stockholm, Sweden, the president of the United States was looking forward to the upcoming meeting of leaders of the economically advanced nations. Entering the airport, surrounded by secret service and advisers, the president was pleased to see the throng of foreign reporters, pushing and shoving to get near the microphone he would use for his press conference. Soon he was behind the microphone.

"Ladies and gentlemen, I would be delighted to answer your questions."

"Mr. President," said a Swedish reporter, "why is it that your nation—among the richest on earth—permits thousands of citizens to be financially broken by medical bills, and millions to worry that the same thing might happen to them?"

For an instant, the president was taken aback. He had not expected this question. But he recovered quickly.

"In the United States we believe that doctors and hospitals should be free to practice medicine without government interference."

"Perhaps so," responded the reporter, "but my question was about paying bills, not the free practice of medicine. Why doesn't every citizen have enough health insurance to prevent financial disaster?"

The president tried to remain unruffled. "We have a fine private health insurance system. Many of our citizens obtain free hospital care."

But the reporter persisted. "I agree that your system takes care of the average person's routine hospital stay. But what about the exceptionally long hospital stay? Or the person with a chronic illness who needs outpatient and home care? And what about the person who is not average—for example, the person who works for a small business that doesn't provide health insurance, or the person who is between jobs and has no coverage? Isn't it true, Mr. President, that your system permits these citizens to suffer an unbearable financial burden due to a medical problem?"

The color was draining from the president's face. He looked around for help. At that moment an aide earned his pay; he calmly took the microphone and said, "The president would love to entertain more questions, but I'm afraid we are already late for our reception. There will be time for questions tomorrow. Thank you very much."

With that, the aide unplugged the microphone and led the president by the arm away from the throng. A half hour later, the president huddled with his advisers in a hotel suite high above Stockholm. He knew he would have to face the reporters again the next day.

"These Scandinavians are so darn righteous about their free medical care. But maybe they're right," he said. "Maybe medical care should be free for everyone. Maybe our government should pay everyone's entire medical bill."

It happened that the president's inner circle included two health economists who spoke up immediately in unison.

"Mr. President, we think that would be a serious mistake. We agree that our lack of universal protection against financial disaster should be remedied. But we shouldn't go to the other extreme. After all, you do remember the Aroman Food Crisis, don't you?"

"The Aroman Food Crisis?" everyone asked all at once. No one had ever heard of it. From their briefcases, both health economists pulled out what appeared to be an old manuscript.

"The hour is late," they said. "Tomorrow morning we can begin to develop a national policy for health insurance. Tonight, relax, and let us read to you from this ancient text. It will prove helpful when we get down to business tomorrow morning."

And so, in unison, they began to read the ancient tale of the Aroman Food Crisis.

The Aroman Food Crisis

An ancient proverb warns, "There is no such thing as a free lunch."
Yet many years ago, in the land of Aroma, there emerged a free-
wheeling, spirited, well-intentioned people, confident that food could,
and should, be free. On a historic day, the Aroman senate rose to its
finest hour and passed the long-awaited Free Food Act. Lest the title
mislead you, in a dramatic amendment before final passage, Senator
Inebriatus, with an impassioned though rambling oration, won the
inclusion of all beverages under the act.

The passage of the Free Food Act (FFA) had been inspired by the
tragic plight of those Aromans with enormous nutritional requirements—fortunately, only a small fraction of the citizenry. But when
the senate's work was done, the act had gone well beyond these citizens. For under the act, food and drink were free for all Aromans. No
longer would any Aroman face even the smallest "barrier" to food
and drink. In other words, no longer would any Aroman have to pay
for what he consumed.

Never again would any Aroman be seen in the marketplace trying
to decide whether a luscious honeydew melon was worth its price.
Nor would the marketplace be degraded by haggling and bargaining
between stubborn buyers and sellers. Now, there would only be smiles.
"Take as much as you want," was the refrain that abounded. "My
price is not for you, my friend, but for our beloved government."
Word spread to other lands about the pleasant vibrations of the Aroman
marketplace.

Yet it was not the gaiety of the marketplace that caused the most
excitement in distant lands. Most wondrous of all were the spectacular public banquets that soon became an everyday feature of
Aroman life. Soon after the FFA was passed, banquet halls sprang
up throughout Aroma. Aromans had always taken a large midday
meal. But now, in retrospect, those meals seemed like a quick snack
by comparison.

Each banquet hall manager boasted of serving only the finest, regardless of cost. When one government regulator asked the manager
of the Endless Lobster, perhaps the most renowned hall in all Aroma,
if perhaps his place was not a bit extravagant, the manager's reply
was angry: "Are you suggesting that I sacrifice quality? I would rather

ties with great success. For example, our water utility company distributes water, and our fuel utility company fuel, to every Aroman home. We regulate the price each charges, and most citizens seem satisfied. We can do the same for our banquet halls."

"What you are forgetting," Economus answered, "is that every Aroman home must pay for the water and fuel it consumes. So people only use what they need, and no more. That's why there is no serious problem. We regulate the price they charge because they're monopolies. But we don't have to regulate utilization, because people do that themselves since they have to pay for whatever they use. That's the crucial difference."

"If we can't regulate banquet halls and marketplace food stands," said Senator Socialus, "then let's take them over and run them ourselves, on behalf of the people." Proposals by Senator Socialus were usually viewed with suspicion by other senators. But now, out of desperation, they listened. "First, we'll put all the chefs and waiters on salary, and stop paying them in proportion to the cost of the food and drink they serve. That will end their incentive to encourage overordering." Many senators nodded, and Socialus continued. "We'll have rational planning of facilities and food production. This is the only way we can solve our problem." While the proposal was radical, many senators seemed to feel it might be the only course left.

But then Economus spoke up. "How will the budget for a banquet hall, and the salaries of those who run them, be determined?"

Socialus answered, "Our government will decide these matters on behalf of the people. We will create a Ministry of Food and Drink to administer this sector of our economy."

Economus responded, "If the Ministry of Food and Drink, rather than consumers, determines the financial success of these providers, then it is the Ministry, not their customers, that they will have an incentive to please."

"But Economus," replied Socialus, "the Ministry will only reward providers who please their customers."

"How will the Ministry know who these are?" Economus asked.

"Well," answered Socialus, "we might take surveys of customers. Or better yet, we could see which providers are greatly in demand and which are not."

"There is a problem with your method," said Economus. "Since

consumers pay nothing, they will prefer the banquet halls and food stands where they are allowed to order as much as they want. The most popular providers will probably be the most wasteful."

"Economus, your trouble is that you always assume financial incentives are everything. Our administrators will realize that they are servants of the people. They will act accordingly, and strive to make the system work for people's needs."

"Socialus, my good friend, I do not think that financial incentives are everything. But I'm afraid I have seen too much in my time to agree with you that they are nothing."

Economus now proceeded to his final point. "I have a final concern about your proposal, Socialus. It applies not only to your plan, however, but to any Free Food Act that would finance all purchases of food and drink through our government budget. Of all the arguments I have made in the course of this senate debate, I believe this one may be the most telling of all. You see, so far my arguments for efficiency have struck some of you as lacking in compassion, and perhaps have therefore left you cold. What I am about to say, however, should at last impress those of you whose sole concern is social justice. For I will now explain why, contrary to the humanitarian intentions of the supporters of the Free Food Act, its effects will be to harm, not help, those Aromans whose needs are greatest—our poor and our elderly."

A silence fell on the senate chamber. Loyal supporters of the Free Food Act, famed for their devotion to programs for the poor and the elderly, turned their attention to Economus for the first time, their countenances reflecting disbelief and concern. The debate had clearly reached its climax, and Economus's ability to defend his assertion seemed likely to decide the outcome.

"As you know," he continued solemnly, "our citizens are unwilling to be taxed without limit. In a dictatorship, where the ruler decides the taxes and people pay them without a whimper for fear of the ruler's wrath, people's sentiments would be no obstacle. But in our Aroman democracy, the situation is very different. We senators must heed the people's wishes, or soon find that others have been elected to replace us.

"Before passing this act, no taxes were needed to finance food and drink. We could therefore devote the taxes we raised to tasks that

only our government can effectively perform—like helping our poor and our elderly. If we must now use taxes to finance all food and drink expenditures—an enormous sum—then we will be unable to continue these vital tasks with the same generosity. It is our poor, and our elderly, who will be harmed the most by the reduction in our other government expenditures. Surely the senators who support full tax financing of food and drink do not realize that its unintended effect would be to shift government spending away from the poor and elderly toward middle- and upper-income Aromans, most of whom can afford to pay their own food bills. I am afraid that the Free Food Act is a sad example of how good intentions can at times lead to harmful results."

A long and sober silence was finally broken by the sound of the gavel of the president of the senate as he adjourned the session.

When the session opened the next day, Naivus was the first to speak. "What shall we do, Economus?" he asked. As usual, he voiced the question then on the minds of most senators.

"My proposal is very simple," answered Economus, "and I am only too glad to set it before you today. First, we must have compassion for those unfortunate Aromans who have enormous nutritional requirements, and simply cannot afford to pay for most of the food and drink they need. We should place a limit, that varies according to the citizen's income, on the out-of-pocket expense that any Aroman must pay for the food and drink he requires. Once a person reaches the limit, our Aroman government should pay the rest. However, the limit would be set high enough so that only those with truly serious needs would reach it. Let's call my proposal *food insurance*.

"It could be easily implemented through our personal income tax. When an Aroman files his annual income tax return, he reports his annual income to our IRS. Using this information, our government can decide each Aroman's maximum burden. For convenience, our government will initially pay every food bill. When an Aroman buys food, he will simply use a government food credit card with his stamped Social Security number. The provider will send the bill to the government for payment. But then the government will immediately bill the Aroman for the amount he owes, according to his income.

"Fortunately, only a relatively small fraction of Aromans are forced

to spend more than that critical fraction of their income on food and drink. Thus, the amount of tax we must raise to help these Aromans is much smaller than the amount required by the Free Food Act. We will therefore be able to continue, and even increase, our government programs to assist the poor and the elderly.

"At the same time, waste and inefficiency will be greatly reduced. Once again, the average Aroman will want to weigh the benefit of anything he orders against its cost. This will automatically limit wasteful over-ordering, without regulation."

Then Economus rose to his conclusion. "Thus, under my simple proposal, no Aroman will be bankrupted by his nutritional requirements; waste and inefficiency will be largely curtailed without cumbersome and costly regulation; and we will conserve scarce tax dollars so that they can be spent to assist the poor and the elderly in other ways."

When Economus had said this, he thanked the senators for listening, and sat down.

Unfortunately, here the tattered manuscript recounting the Aroman food crisis becomes illegible, and to this day we do not know whether the Aroman senate followed his advice.

Having completed his recounting of the ancient story, the president's health economists looked up. Everyone was sound asleep except the president, who nodded his head drowsily with a sign of approval. Within seconds, he too was asleep.

The next morning, the president asked what to do about medical coverage. He turned to his two health economists, and as usual, they spoke in unison. "We agree about the central lesson of Aroman Food Crisis—that it's a mistake to make something completely free. But we don't agree about what to do about health insurance. We each have a different plan. So from now on, we will no longer speak in unison."

"No longer in unison?" the president asked with surprise.

"We're afraid not," they said, still speaking in unison.

"My plan is called Health Card," said one health economist, speaking alone.

"And my plan is called a Health Insurance Tax Credit," said the other health economist, also speaking alone.

"This is going to make it hard for me to arrive at a decision," said the president glumly.

"We're sorry," they said in unison.

"No, I'm the one who's sorry," said the president. Addressing one health economist, the president said, "You present your Health Card plan," and then turning to the other health economist he said, "and then you present your Health Insurance Tax Credit plan."

"We will," they said in unison.

Health Card

"Mr. President," said the first health economist, "I urge you to propose a new plan called *Health Card*. Health Card would utilize a government issued health credit card and the income tax return data of the Internal Revenue Service. It would therefore depend crucially on modern computer technology that would have been unfeasible until recently. Health Card would achieve automatic universal coverage regardless of employment or health status, and cost containment through *equitable* patient cost sharing."

"But how would Health Card work in practice?" asked the president.

The economist replied, "Every household (regardless of employment or health status) would receive a health credit card issued by the federal government. The household would use its health card for medical care the way it uses a MasterCard or Visa card for other goods and services. The medical provider would send a patient's bill to the government's agent (a private credit card company such as Visa or MasterCard, or an insurance company such as Blue Cross/ Blue Shield), who would fully pay the provider's bill using government funds. The government's agent would then bill the household for a percentage of its medical bill. This percentage would be scaled to the household's income as reported on its most recent federal income tax return. Once the household's financial burden reaches a designated percentage of its income, it would not be billed again that year, so Health Card would limit every household's financial burden to its ability to pay.

"Like MasterCard or Visa, the government's agent would fully pay the medical provider's bill, and then bill the household for a percentage of the bill. The percentage might be 5 percent for a low-income household and 25 percent for a high-income household. But the low-income household would not be billed for more than 1 per-

cent of its income and the high-income household would not be billed for more than 5 percent of its income. A middle-income household with an income of $60,000 might be billed 15 percent of its medical bill until its annual medical bill reaches $12,000 and the household's burden reaches $1,800 (15 percent of $12,000) which is 3 percent of its income (3 percent of $60,000); that household would not be billed for any additional medical bills incurred that year so its maximum burden would be 3 percent of its income.

"A Health Card table would be included in the 1040 federal income tax booklet along with the usual tax rate tables. If a household had an unusually high burden last year, this year's maximum burden and cost-sharing rate would be reduced in order to give further protection to any household with a chronic high-cost medical problem."

"How ironic," noted the president, "that Health Card's fairness depends on using tax return data from the Internal Revenue Service, surely not our most popular government agency. But Health Card will involve new taxes. How am I going to persuade people to accept new taxes?"

"It won't be as hard as you fear," answered the health economist. "Under Health Card, employers would send checks to the government ('taxes') instead of to private insurance companies ('premiums'). The average employer and employee would hardly notice this change. Employers would simply send comparable checks to a different address. Of course, additional revenue would be needed to cover those currently uninsured. Thus, the current Medicare payroll tax would need to be raised a few percentage points and be renamed the Health Card payroll tax."

"But what's to prevent a household from obtaining private insurance to cover the fraction that the government won't pay, so that medical care stays free to patients?" asked the president.

"Good question," replied the economist. "Fortunately, the answer is simple. Under my proposal, the government won't pay anything if the person receives reimbursement from private insurance."

"What do you mean?" said the president.

"Simply this. When the provider sends the bill to the government, it must indicate that it is not submitting the bill to a private insurer, and must enclose the patient's signature making the same pledge."

"But then," asked the president, "why would a household want to buy

The Health Insurance Tax Credit

"So now tell me about your Health Insurance Tax Credit," said the president to the second health economist.

"Gladly," replied the economist. "The Health Insurance Tax Credit (HITC) would enable everyone to obtain private health insurance, either through the person's employer or directly from an insurance company."

"How?" asked the president.

"Consider a person whose employer does not provide health insurance. If the person buys her own health insurance from an insurance company, she would be able to claim a new health insurance tax credit on her April 15 personal income tax. For example, a single person might claim a $2,000 tax credit."

"How would this work?" asked the president.

"Suppose she would have owed the government $3,000 in tax. Then with the $2,000 tax credit, she only has to pay the government $1,000."

"But suppose," said the president, "that she would have owed only $1,000 in tax?"

"Then," replied the economist, "she would file her tax return documenting her purchase of health insurance, and the government would send her a check for $1,000. She would get a "refund" of $1,000. So the health insurance tax credit would be *refundable*."

"I see," said the president. "What if she has a husband and two children, and they file a joint tax return?"

"Then her family might be entitled to a $5,000 tax credit—$2,000 for her husband, and $500 for each child."

"Would every family, regardless of income, get the same tax credit?" asked the president.

"No," answered the economist. "The higher the income of the family, the less tax credit it needs to afford health insurance. So under my HITC plan, the credit would phase down as family income rises. For example, a high-income family of four might receive only a $2,000 tax credit, but a low-income family of four might receive $8,000. Hence the average tax credit for a family might be $5,000. Each family's April 15 income tax return would contain a table indicating the amount of credit to which it was entitled."

"That makes sense," replied the president. "But I'm worried about something. Consider an employer currently providing health insurance. Might not that employer say to his employees, 'Look, you can now get a tax credit if you buy health insurance on your own. So I'm not going to offer you health insurance any more; instead, I'll give you a higher wage, and then it's up to you find your own insurance.'"

"I'm a step ahead of you," replied the economist. "I have two methods of making sure employers offer insurance to their employees. First, my HITC plan would provide the same tax credit to a family if it obtains insurance from its employer. Second, my HITC plan would require employers to offer health insurance to their employees."

"You *are* a step ahead," said the president. "But I have a question about each of your two methods. First, is it fair to give the employee a tax credit when the employer pays for the health insurance?"

"In fact, it *is* fair, because even when the employer pays the insurance company, it is the employee who indirectly pays for the insurance."

"I don't understand," said the president.

"Let me explain," said the economist. "If an employer pays an insurance company to provide health coverage, then the employer won't be able to pay as much in cash wages and salaries. For example, if the employer pays $10,000 per employee for family health insurance, the employer will probably pay about $10,000 less per employee in cash wages or salaries. So employees indirectly pay for employer-provided health insurance by receiving lower cash wages and salaries than they otherwise would."

"That seems correct, now that I think about it," said the president.

"So," continued the economist, "imagine two identical workers. The first works for an employer who pays a cash salary of $50,000 but no health insurance, and the second works for an employer who pays cash salary of $40,000 but pays $10,000 for the employee's health insurance. Suppose the first worker buys health insurance on her own for $10,000, and that with an income of $50,000 that worker can claim a tax credit of $5,000. Then after buying health insurance for $10,000 and getting the $5,000 tax credit, the first worker has $45,000 left. If the second worker were not also given a $5,000 tax credit, the second worker would have only $40,000 left. But if the second worker is also given a $5,000 tax credit, then that worker will have $45,000 left, just like the first worker. Since both pay, directly

or indirectly, $10,000 for health insurance, to be fair both should get the same $5,000 tax credit."

"I see," said the president. "It *is* fair. Of course, it will be more expensive for the government. Any suggestions about how to get some money to pay some of the expense?"

"Yes," replied the economist. "Fairness also requires that both of these workers pay income tax on an income of $50,000. But under current law, the first worker pays tax on $50,000, but the second only on $40,000, because under current law, only cash wages and salaries are subject to personal income tax every April 15."

"Can that be fixed?" asked the president.

"Easily," said the economist. "Just include an employer's expense on health insurance for an employee in the income of the employee that is subject to April 15 tax. Both workers would receive a W-2 form from the employer at year's end indicating $50,000 of income subject to tax; the first worker's W-2 form would note a cash salary of $50,000, and the second worker's W-2 form would note a cash salary of $40,000, and a health insurance benefit of $10,000. So now the government would collect tax on $50,000 of income from the second worker instead of on only $40,000, so this would raise some more tax revenue for the government, and help pay for the new tax credit."

"Excellent," exclaimed the president. "But now I have a question about your second method—requiring employers to offer health insurance. In the past, a proposal to require employers to buy health insurance for their employees has met stiff political resistance from many small business managers who claim they can't afford to buy such insurance."

"My proposal is different," said the economist. "It does not require employers to *buy* insurance. It only requires them to *offer* insurance. Each employer would be required to arrange with a private insurance company to offer to sell health insurance to its employees. The employer would *not* be required to pay any of the premium. Of course, employers would be free to pay part or all of the premiums."

"But does the offer really help the employees if they must pay the whole premium?" asked the president.

"Yes, for two reasons," replied the economist. "First, the employer will usually be able to get a substantial group discount from an insur-

ance company—a lower premium per employee than if the employee bought insurance on his own. Second, the employer will be providing employees the service of searching the insurance market for a satisfactory policy and scrutinizing the details of any insurance policy it offers; most employers will do a decent job because they will want to avoid employee dissatisfaction with their insurance policy—dissatisfaction that can lower employee morale and productivity."

"I see," said the president. "Requiring an employer to offer, but not necessarily buy, insurance sounds like a nice compromise."

"Now," said the economist, "let's turn to another crucial advantage of the HITC: It will cause people to weigh cost against benefit."

"I don't see why," said the president.

"Let me explain," replied the economist. "A family of a particular income level would receive a particular tax credit regardless of its premium. For example, a $50,000 family might receive a tax credit of $5,000; if its insurance premium is $10,000, it would bear a burden of $5,000, but if its premium is $1,000 less, $9,000, its burden would be $1,000 less, $4,000. So it would gain $1,000 of cash by reducing its premium $1,000. It can reduce its premium by selecting an insurance plan that has patient cost-sharing for moderate bills. Thus, it is likely that under HITC, many people would opt for insurance plans with moderate premiums and moderate patient cost-sharing. But this means that they would usually bear some of the cost of any medical care they order, so they would usually have an incentive to weigh cost against benefit."

"But," asked the president, "what if their employer buys the insurance?"

"Same thing," replied the economist. "Their employer would know that their employee's tax credit stays the same regardless of the premium. If the employer buys a plan for $1,000 less, the employer would be able to pay $1,000 more in cash wages, and there would be no change in the employee's tax credit. So it is likely that employers would opt for insurance plans with moderate premiums and moderate patient cost-sharing."

"I have a question," said the president. "Two families of four each with $50,000 of income would receive the same tax credit, say, $5,000. But suppose one family is old and one is young. Or one is unhealthy and the other is healthy. Insurance companies would charge the old

or unhealthy family a high premium, and the young or healthy family a low premium. So is it fair to give all $50,000-income families the same $5,000 tax credit?"

"That's a good point," said the economist. "It would indeed be fairer to adjust the tax credit not only by income, but also by age and health, provided poor health is due to bad luck, not bad behavior (such as chain smoking). Adjustment by age would be easy to implement. Adjustment by health, however, would be harder. It would be administratively unfeasible to have every family obtain an assessment of its 'health.' In a small number of extreme cases, however, it might be possible for a family to qualify for an adjustment based on a documented diagnosis by a physician."

"That sounds better," said the president. "But what happens if a family's health is so poor that no private insurance company is willing to cover it at a reasonable premium?"

"Good question," said the economist. "I think the HITC should be supplemented by a government guarantee of last-resort insurance."

"What do you mean?" asked the president.

"The government should stand ready to offer insurance at an affordable premium to any family either rejected, or charged an exorbitant premium, by private insurance companies.

"Would the government be the insurer?" asked the president.

"It might be, but it wouldn't have to be. Instead, the government could offer to pay private insurance companies a fee for agreeing to insure unhealthy families for a reasonable premium. Private insurance companies in each region would bid competitively to obtain the government contract for last-resort insurance."

"Another question," said the president. "Suppose someone loses her job and thereby loses her health insurance. What then?"

"Last-resort insurance at a reasonable premium should be guaranteed for anyone between jobs," replied the economist.

"Something still bothers me," said the president. "Some people are always going to take their chance going without insurance and spending their money on other things. Then when they get hit with a huge hospital bill, they can't pay it, so the hospital charges everyone with insurance more to make up for it. So people who are insured end up paying for people who aren't insured. That's unfair."

"I agree," said the economist.

"It's unfair," continued the president, "to let some people be freeriders at the expense of everyone else. We're not going to turn them away from the hospital when they need it. So it's only fair to make them pay for insurance like everybody else."

"We require people to obtain auto insurance," said the economist. "We could do the same for health insurance."

"I think we should," said the president.

"So let me summarize my plan," said the economist. "It provides a health insurance tax credit to everyone who obtains insurance, offers last-resort insurance at a reasonable premium, requires every employer to offer health insurance, and requires every person to obtain health insurance."

"Sounds just right to me," said the president.

The Tax Credit for Working Families and Health Card for Retirees

The president spoke to his two health economists. "You have each made a persuasive case for your own plan. Both plans have great merit. I am persuaded that the tax credit plan is best for working families because the majority currently have satisfactory private health insurance through their employer, and the tax credit will let them keep their private insurance while also extending it to other working-age families that currently do not have it. For working-age families, the tax credit must be supplemented by providing last-resort insurance at a reasonable premium, requiring every employer to offer health insurance, and requiring every person to obtain health insurance.

"However," continued the president, "most retirees cannot get private insurance through an employer. Currently, they get their basic insurance from the government through Medicare, sometimes supplemented by private insurance that covers what Medicare does not. For retirees, I think Medicare should be reformed into MediCard—Health Card for retirees. MediCard would cover the same services Medicare covers plus prescription drugs. Just like Health Card, MediCard would bill patients a small percentage that is scaled to their income."

The two health economists looked at each other, then at the president, and smiled.

"That's a nice compromise," the two health economists said in unison.

monopoly feels little pressure to perform well, because it has a captive audience. Consumers must buy its product. They have nowhere else to turn, no matter how poor a job the monopoly does. And the monopoly knows it. So the monopoly slacks off. It retains unproductive workers. It innovates slowly. And it responds sluggishly to consumer complaints."

"I know, I know," exclaimed XT. "What a contrast between your economic system and the communist system of the old Soviet Union. I'll never forget my visit to Moscow in the 1980s, before the fall of communism. What long lines there were at the Moscow department store! The poor Russians! How they grumbled. And how naive I was. I thought they must be willing to wait this long because the products were so wonderful. When I finally reached the front of the line, I couldn't believe it. What shoddy merchandise. But the poor devils simply had no choice."

"But XT," said his host, "you should have expected it. The Soviet economy consisted of monopolies. Oh sure, they were public monopolies, run by the state, supposedly dedicated to serving the people. But a monopoly is a monopoly. Whether it's public or private hardly matters. When consumers have no choice, when producers feel no pressure from competition, the result is the same: poor performance, shoddy goods and services."

"How astonishing," exclaimed XT, "but you are right. I have seen it with my own eyes. It's not even close. Competition dramatically outperforms monopoly. You should be quite proud of your little mechanism called competition."

"We are," replied his American host, beaming, "we are."

XT's Confusion

"But," continued XT, "there is something that confuses me. I hope you won't take offense at the questions I am about to ask."

"Not at all," replied his host with confidence. "Fire away."

"Wouldn't you agree that it is more important to get excellent performance from producers of education than from producers of, say, furniture?"

"I certainly would agree," replied his host.

"So," asked XT cautiously, "please don't take offense, but why do you use local monopolies to produce education?"

The host felt struck by a bolt from the blue. He simply hadn't seen it coming. For a moment he was speechless.

"After all," continued XT, "I would have thought that education would be the most important place to use your astonishing mechanism of competition to get the best performance possible. And to my surprise, I find this is one of the few areas where you tolerate monopoly. I'm afraid I don't understand."

"But," said his host, "when it comes to education, we believe that each community should form a single enterprise, called a public school. We believe that the best minds in each community should come together in this enterprise and plot its course. Then everyone should cooperate and get the job done. And by all means, we shouldn't let other enterprises proliferate and wastefully duplicate effort. One cooperative effort, working in harmony for a single goal, is surely better than many enterprises working at cross purposes, engaged in petty rivalry with each other."

"Why," exclaimed XT, "that's the way everyone on my planet thinks. Are you sure you haven't visited there? That's what I thought until I came to Earth and discovered your marvelous invention, competition."

"But education is different," said his host.

"Yes," said XT, "it's more important. All the more reason to use your best weapon, competition. Please forgive me if I quote you. Didn't you tell me earlier that a monopoly feels little pressure to perform well because it has a captive audience? Consumers must buy its product. They have nowhere else to turn, no matter how poor a job it does. So the monopoly slacks off. It retains unproductive workers. It innovates slowly. And it responds sluggishly to consumer complaints. Did I dream it, or weren't these your very words?"

"Yes, yes, they were," muttered his host. "But our local monopolies are dedicated to serving the community."

"Yes," replied XT, "but didn't you also say that a monopoly is a monopoly? Whether it's public or private hardly matters. When consumers have no choice, when producers feel no pressure from competition, the result is the same: poor performance, shoddy goods and services."

"Maybe I did say that," admitted the host reluctantly. "But our local monopolies are different. Our teachers and principals are very dedicated. And the consumers—parents—form associations that monitor the school. They do apply pressure. And there is also pressure from parents through the election of school board members."

"Please explain, Economas," said the surprised host.

"Even though public schools are free and private schools charge tuition, in affluent neighborhoods the school board, administrators, and teachers all know that if they do a poor job, a significant number of parents will switch their children to private schools. The fall in enrollments will reduce tax revenues sent to the public schools, and this will eventually reduce the salaries and employment of administrators and teachers. So private schools do exert competitive pressure on public schools in affluent neighborhoods. Moreover, affluent educated parents are active in monitoring school performance and working with school administrators and teachers. So affluent parents pressure their public schools in two ways: first, they have the ability to switch to a private school; and second, they directly interact with administrators and teachers."

"Wouldn't it be better still if there were a level playing field between public and private schools, with all schools charging tuition, and none directly receiving tax revenue?" asked XT.

"I disagree," replied Economas. "Let me explain. To my children, it matters a great deal which school their friends attend. Economists call this a *consumption externality.* If my eight-year-old son's friend gets a birthday present my son wants, he may be jealous for a day, but he'll quickly get over it. But if his friend gets accepted at one school, and he gets rejected, it would take a long time for him to get over it."

"I've heard that even high school seniors need some time to get over such an experience concerning college admissions," admitted XT.

"Parents and children," continued Economas, "care very much about who else attends their school. But for just this reason, competing schools will be selective in their admissions. They will act like colleges. They will have admission tests, accepting some, rejecting others. Just as colleges get ranked, elementary schools will get ranked. One eight-year old will get into a high-ranked school while her best friend is rejected."

"But couldn't the government prohibit elementary schools from being selective in their admissions?" asked XT.

"It could try," said Economas. "But I'm not very confident it would work. This would be a major change for our private schools that have always sought information about applicants. They might

prefer to stay out of the voucher plan rather than submit to blind admissions."

"It won't be easy to enforce blind admissions," said the host. "Schools will resist simply being assigned students by lottery. If they are allowed to interview the applicant's family, it will be easy to obtain information about educational background and occupation. Without a lottery, it seems likely that school rankings will develop, and rejections will split friends."

"Even if blind admissions could be enforced," said Economas, "there is still a problem. Today, youngsters in the same neighborhood usually attend the same public school. Friends can be sure they can go to the same school, and when friendships change, it will be among youngsters in the same neighborhood. These friendships often lead to friendship among parents and build a sense of community. But now suppose that neighborhood friends are admitted randomly to different schools. It's not as bad if the choice is random rather than based on test scores or grades. But it still splits friendships."

"I see your point," admitted XT.

"Economists call this a *consumption externality* because if neighborhood children are split among a substantial number of private schools, many children will be adversely affected due to rejections, split friendships, and a weakening of neighborhood ties. The externality implies that there should be a strong financial bias in favor of the neighborhood public school. A family should be able to switch to a private school—indeed, this threat provides some competitive pressure on the public schools—but the family should have to bear a significant financial burden so that switching is the exception rather than the rule."

"Is there any other externality that warrants favoring public schools?" asked XT.

"I think there is," replied Economas. "When children of different incomes, races, religions, ethnicities, genders, and disabilities attend the same public school, most learn to interact and work together, and this experience carries over when they enter the workplace and residential community. Employers may be more willing to hire and promote according to skill and merit, rather than according to race, religion, ethnicity, gender, or disability. There may be greater toler-

ance in political as well as economic life. While some private schools achieve this mix, many do not. Public schools are supposed to plan sufficient capacity so they can accept all applicants, and to treat all students as first class citizens. This philosophy may stay with many students when they become adults, thereby strengthening our nation's commitment to tolerance."

Economas's Targeted Scholarship Plan

"Does this mean," asked the host, "that you are against a voucher plan that applies to all families and all schools?"

"That's correct," replied Economas. "I favor maintaining a strong bias in favor of neighborhood public schools by keeping them free, financed by taxes. Affluent families already have the option of choosing a private school because they can afford private school tuition, and this option already applies some competitive pressure to their public schools. So I think we already have it just about right for the affluent. But I agree that there is not enough choice for nonaffluent families, and not enough competitive pressure being applied to their public schools."

"So what do you propose?" asked the host.

"I propose a targeted scholarship plan," replied Economas.

"How would it work?" asked the host.

"First," replied Economas, "public schools would remain free, financed by taxes; a family's scholarship would cover only a percentage of private school tuition and the family would be required to bear a significant financial burden if it chooses a private rather than a public school; the percentage should be higher for low-income families but should always be less than 100 percent. Second, the scholarship would be targeted on nonaffluent families; the scholarship amount would phase down to zero as family income rises from low to middle income. Third, the scholarship could only be used at schools that do not discriminate on the basis of income, race, religion, ethnicity, gender, or disability. The scholarship might be used at an accredited school with a religious affiliation provided the school does not discriminate on the basis of religion in its enrollment, and provided the Supreme Court concludes that its use does not violate the constitutional separation of church and state."

"But how would the government know each family's income?" asked the host.

"Every family," replied Economas, "seeking a scholarship would file a federal income tax return, reporting its income. Most families already file, but some would now file just to obtain the scholarship. Based on the information it receives on the return, the U.S. Treasury would write the family a check for an amount of scholarship scaled to its income."

"Your targeted scholarship plan differs significantly from XT's voucher plan," said the host.

"True," replied Economas, "but like XT's plan, it would still apply some competitive pressure to public schools that enroll primarily nonaffluent students, and give some choice to nonaffluent parents who are very dissatisfied with their public school. After all, affluent public schools are already subject to some competitive pressure, and affluent parents already have some choice. As an economist, I want to favor neighborhood public schools over private schools because of the 'consumption externalities' I explained earlier. But as an economist, I also want some competitive pressure applied to all public schools. Today, poor families are trapped—most can't afford to switch to any private school, no matter how bad the situation is at their public school, and the public school knows it. A targeted scholarship would make switching a real possibility, and create real competitive pressure on these public schools for the first time."

"But," said the host, "I wonder whether the most concerned parents will switch their children, while children with unconcerned parents will stay in bad public schools. These children will lose the positive influence of the better students and will be even worse off."

"I admit this is possible," replied Economas. "But don't forget a key point: The new competitive pressure on public schools in poor neighborhoods should improve public school performance. For the first time, administrators and teachers at these schools would worry about losing their jobs as a result of a fall in enrollment. The threat of concerned parents switching should improve public school performance and thereby reduce the actual number who switch. So I think it is more likely that children of unconcerned parents will be better off."

"But in some places the public school will stay bad," replied the

host. "The best students will switch, and children with unconcerned parents will be worse off. Maybe we need to keep the best students in the bad school for the sake of the other children."

"Would you leave your child in a bad school for the sake of children whose parents don't care?" asked Economas.

"No," replied the host.

"Then I don't think it's fair for you to ask concerned poor parents to sacrifice their own children when you wouldn't sacrifice yours." said Economas. "Nonaffluent families deserve some choice and some competitive pressure on their public schools—something affluent families already have and take for granted."

Reducing Inequality in Public School Spending per Pupil

"Something else is crucial for improving the education of the non-affluent," said Economas. "We need to reduce inequality in public school spending per pupil."

"I don't understand," said XT.

"In your fascination with competition, XT, you have overlooked a simple point," said Economas. "Affluent children get much more spending per pupil than nonaffluent children."

"But," said the host, "don't we try to reduce this inequality by state and federal aid formulas? I thought that states give more aid per pupil to poor school districts than to affluent school districts, and that most federal aid is targeted on poor school districts."

"That's true," replied Economas, "but great inequality still remains. On average, affluent districts spend about 50 percent more per pupil than poor districts."

"Fifty percent is an enormous difference," admitted the host. "But is expenditure per pupil really that important for educational quality? Isn't the education and income of the families of a school's children more important?"

"That may be," replied Economas. "But the affluent already have their children in a school where parents have high education and income, so if expenditure per pupil doesn't matter, why do they choose to spend 50 percent more than poor districts?"

"Good question," admitted the host.

"The affluent recognize," continued Economas, "that higher expenditure per pupil buys higher-quality teachers, newer textbooks, better-equipped science labs, newer and better computers, and fewer students per teacher. It would be very odd if greater expenditure didn't yield higher educational quality. After all, it does for appliances, furniture, houses, cars, roads and bridges, hospitals, police, the military, and countless other goods and services. Why shouldn't it yield higher quality for education?"

"But I know some affluent people," said the host, "who insist that expenditure per pupil doesn't matter."

"Watch what they do," said Economas, "not what they say. They are free to spend the same as poor districts. But they don't. On average, they choose to spend 50 percent more per pupil."

"Maybe what they mean," said the host, "is that higher expenditure per pupil won't do much good as long as a school fails to maintain discipline in the classroom."

"That's where my targeted scholarship plan should help," replied Economas. "If families can afford to switch to private schools, there will be competitive pressure on public schools to maintain discipline. With better discipline in the classroom, greater expenditure per pupil will surely improve educational quality."

"But it's more than that," said the host. "Greater spending per pupil may not help much as long as drugs and crime are rampant in the neighborhood."

"I agree," said Economas. "Reducing drugs and crime should be a top priority. Once again, that requires higher expenditure on police, courts, and prisons, and much of this money must come from taxes raised outside the poor neighborhoods. The aim should be to help law-abiding, working parents in those neighborhoods who are concerned about their children's education and safety."

"But which level of government should try to reduce inequality in public school spending?" asked the host.

"Both the state and federal government," replied Economas. "Both already provide public school aid that favors low-income districts. But significant inequality remains, because federal taxes finance less than 10 percent of elementary and secondary education, while local taxes finance nearly 50 percent. The strategy should be to raise state and especially federal tax shares while reducing local

tax shares for public schools. Perhaps earmarking certain state and federal taxes for public school aid would be helpful in obtaining political support."

A Computer Tax Credit for Low-Income Families

"I have a different point," said the host. "The inequality problem is growing more severe because of the personal computer and the Internet. Affluent parents spend much more to make sure their children have a high-quality personal computer and access to the Internet in their own home."

"You're right," said Economas. "Suppose a student is assigned to write a history report. In the affluent neighborhood, the student sits down after dinner at her computer with her own encyclopedia and perhaps her own history book. With her computer, she searches the Internet to find additional information. Then she types her first draft on the computer. Later, she works on the draft to revise it and uses the computer's spell check. But in the poor neighborhood, the student has no encyclopedia, no history book, and no computer. Her only chance is the school or public library. If she can get access to an encyclopedia, a history book, and a computer with access to the Internet, she is in luck. Otherwise, she's in trouble. Even if she can use the library's encyclopedia, without access to a computer she must write her first draft by hand. Then she must write the final draft from scratch. She must use a dictionary rather than a computer spell check. Finally, she probably must do her work before dinner, because the school or public library is probably closed in the evening, and even if it were open, it might not be safe to come home in the dark."

"What can we do about this?" asked the host.

"We can enact a computer tax credit for low-income families on the federal income tax," replied Economas. "To claim the credit, the family would include the receipt documenting its purchase with its tax return. The credit would reimburse the family for a percentage of the computer price up to a maximum. The credit would phase down to zero as family income rises from low to middle income. The credit would be 'refundable,' so that if a poor family's computer credit exceeds the tax it owes, it would get a 'refund' from the

IRS. For example, if its computer credit is $900 and its tax is zero, the Internal Revenue Service would send it a check for $900, just as it sends refund checks to many households. Remember, the computer tax credit never completely removes the financial burden on the family—the percentage reimbursed is always less than 100 percent. So most low-income families will try to take good care of their computers. Of course, here's one more reason to reduce crime in poor neighborhoods."

A Compromise

"One final point," said the host. "Some citizens favor voucher plans but oppose reducing the inequality in public school spending, while other citizens favor just the reverse. But you favor both a targeted scholarship plan and reducing inequality in public school spending."

"That's right," said Economas. "We need both. We need a targeted scholarship plan for nonaffluent families to give some choice to these families and to provide some competitive pressure on their public schools. Affluent families already have some choice and competitive pressure, and take it for granted. At the same time, we need to reduce inequality in public school spending by shifting more of the financing away from local governments to the state and federal governments that can and do redistribute funds from affluent to nonaffluent districts."

"You're advocating a political compromise between the two sides," said the host.

"Exactly," said Economas. "I'm appealing to public school advocates to accept a targeted scholarship plan in return for more generous state and federal funding of public schools. And I'm appealing to voucher advocates to accept more state and federal funding of public schools and a targeted scholarship plan. To reassure both sides, they should be enacted as part of the same package. One hopes both sides will support the computer tax credit for nonaffluent families."

"What do you think, XT?" asked the host.

"I've learned," XT replied, "not to get too excited about any mechanism, however fascinating, especially when one doesn't know certain complexities of the inhabitants of a strange planet. Economas has persuaded me that unrestricted competition is unwise for elemen-

tary and secondary schools. However, I am pleased that Economas agrees that some competitive pressure does have a role to play in improving public school performance, and that a targeted scholarship plan would be worth enacting. I now see that reducing inequality of public school spending per pupil deserves high priority and that a computer tax credit would be warranted for nonaffluent families. Incidentally, I've become fascinated with your computers and your Internet. We don't have them on my planet. Any chance of an interplanetary connection?"

13

Poverty

We've got two economic problems at the bottom of our economy. The first problem is this: There are adults who work full time, but their low education and/or skill results in a low annual income. This has always been a problem, and it has intensified during the past two decades because wage inequality has been increasing in both the United States and most other economically advanced countries. Most economists cite two reasons for rising wage inequality: technological change, and greater international trade with low-wage countries. It is not immediately obvious that technological change must widen wage inequality between highly educated and less educated persons. After all, new machinery generally raises the productivity and wage of all workers, and over the past two centuries many new machines especially raised the productivity of workers with little education. It appears, however, that recent technological change, often involving computers, has favored highly educated workers more than less educated workers. Moreover, advances in telecommunications and transportation have increased international trade with low-wage countries, and this tends to put downward pressure on the wages of less educated workers in economically advanced countries.

Here's the second problem: too many less educated persons, both adults and teenagers, rely on welfare rather than work. Some don't

want to work for low wages, while others want to work but often can't find an employer willing to hire them. There's a lot of disagreement about where the fault lies, but there's widespread agreement that work is better than welfare.

This chapter proposes an expansion of the earned income tax credit to treat the first problem, and the provision of last-resort low-wage jobs to treat the second.

The Earned Income Tax Credit

People who work full time for low annual incomes are assisted by the earned income tax credit (EITC). If you've never heard of the EITC, you're not alone. The minimum wage and welfare are much more famous. But the EITC is much better than raising the minimum wage or giving welfare to persons able to work. Let me explain why.

The EITC was enacted by Congress in the mid-1970s, and greatly expanded in the 1990s. It's a tax credit on the federal income tax. In contrast to welfare, it's available only to people who actually work. No labor earnings, no EITC. Imagine yours is a low-earning household with two children. Your EITC is based on your household's total labor earnings. As shown in Figure 13.1, for each $100 you earn, the government adds $40, until at $10,000 of earnings, the government supplement reaches $4,000 (these are round numbers that are approximately correct). Then, to avoid paying supplements to everyone, the phase-out begins. For each additional $100 you earn beyond $10,000, the supplement is cut $20. When your total income reaches $30,000, the supplement has been cut to zero. So if your household earns less than $30,000 of total income, you're entitled to some supplement from the government. The maximum EITC credit and dollar thresholds are automatically adjusted for inflation.

But what if your earnings are so low that you don't owe any federal income tax? You still can get the full EITC supplement. EITC is a *refundable* tax credit. This means that if your EITC is greater than the tax you owe, the government will write you a check for the difference. For several years after the EITC was enacted, many low-income workers had never heard of it. But since its expansion in the 1990s, the word has spread in most low-income neighborhoods. Today, most (though not all) low-income workers have heard of the EITC, and most seek assistance

Figure 13.1 **Earned Income Tax Credit**

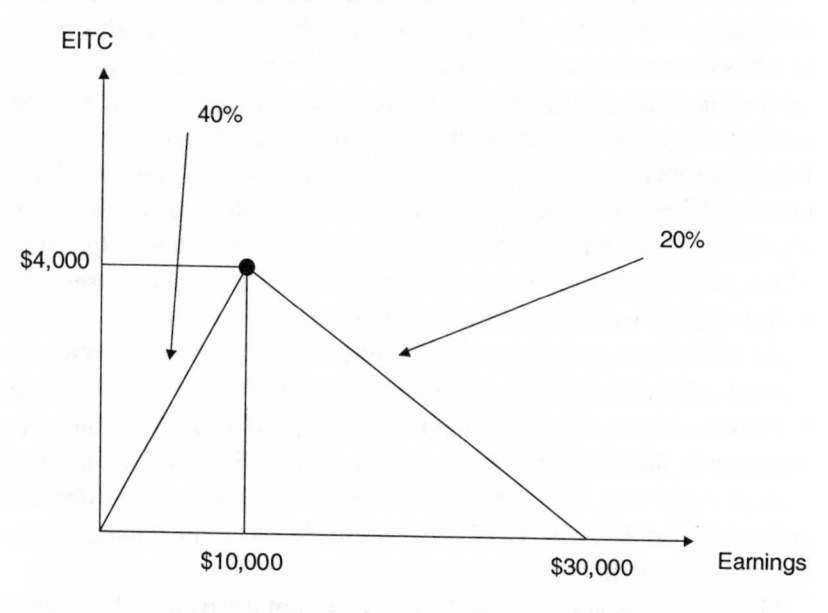

from tax preparers to help them file a tax return so they obtain the credit.

Why is the EITC better than raising the minimum wage? Because raising the minimum wage probably reduces employment. An employer may find it profitable to hire a less educated person if he can pay him $6 an hour, but not if he must pay him $8. You can pontificate all day about what employers should do. But the hard fact is that employers will probably offer fewer jobs the higher the minimum wage is set.

The EITC response is this: Let employers offer more jobs at low wages. Then provide a supplement for workers. This way we get the best of both worlds—more jobs, and higher incomes (wages plus the EITC supplement) for workers.

The EITC is also better than the minimum wage because it is targeted. Raise the minimum wage and you raise the wage of teenagers from affluent households, while you reduce jobs for low-skilled heads of households. Raise the EITC, and you raise the money that goes only to households with low annual earnings.

The EITC is better than welfare because it sends the right message. No work, no assistance. If you are capable of work, the message should be: Forget welfare. You can earn an EITC supplement by working.

vate firms to run the projects. But the government must finance the projects. It must make good on its guarantee to provide anyone willing to work with a low-wage job. Only with that guarantee is it acceptable to end welfare for anyone able to work.

But consider a young unmarried mother who refuses to work, thereby subjecting her children to starvation. True, we can't let her children starve. But we can judge her guilty of child neglect. There must be due process. She must be asked, "If you can't find a job, there's a place to go to get one: the local jobs center. Will you do it?" As long as she goes and does the job she is given, her children will not starve. If she is not willing to take the job, then we have no choice but to judge her guilty of child neglect and unfit to have custody of the children.

Now, obviously, it is usually best for children to stay with their parents. And in most cases, the young unmarried mother will go to work to keep her children. But in some cases, until drug or alcohol abuse is overcome, it may be better for the children to have another home.

What about the unmarried father? He can get a last-resort job too. Suppose he refuses? Quite often the young mother knows who the young father is. If she can no longer get welfare but has to work to feed her children, she may try to get the father to take a job and help support his children. She may be willing to name him, so the government can pursue him for child support.

But suppose none of this works. Suppose teenage girls see that if they become pregnant without marriage, they face work, not welfare. Suppose they see that the young father often gets off, and the young mother must work to keep her children. It's likely that more teenage girls will try harder to not get pregnant.

Of course, low-wage jobs have trouble competing with drug dealing or stealing. So just as the government must provide last-resort jobs, it must be tougher on drugs and crime. The carrot and stick must both be used. The aim must be to create an environment where a teenage boy says, "If I do the right things—go to school regularly, do my homework, stay away from drugs and crime, and don't father a child until I'm married and can support my child—then I will be rewarded; but if I do the wrong things, I'll be in trouble." Similarly, we must create an environment where a teenage girl says, "If I do the

right things—go to school regularly, do my homework, stay away from drugs and crime, and don't get pregnant until I'm married and can support my child—then I will be rewarded; but if I do the wrong things, I'll be in trouble."

It sounds conservative to say: No welfare, and be tougher on crime and drugs. It sounds liberal to say: Government will provide last-resort low-wage jobs. But we need both conservative and liberal wisdom to make progress at the bottom of our economy.

About the Author

Laurence Seidman is Chaplin Tyler Professor of Economics at the University of Delaware. He previously taught at the University of Pennsylvania and Swarthmore College. He has won the teaching award of the College of Business and Economics at the University of Delaware, and has been teaching a one-semester introduction to economics for many years. He is the author of *Automatic Fiscal Policies to Combat Recessions* (2003), *Funding Social Security: A Strategic Alternative* (1999), *The USA Tax: A Progressive Consumption Tax* (1997), and (with co-author Saul Hoffman) *Helping Working Families: The Earned Income Tax Credit* (2003). He has published economics articles in the *American Economic Review, Journal of Political Economy, Review of Economics and Statistics, Journal of Public Economics, National Tax Journal, Brookings Papers on Economic Activity, Public Finance Review, Southern Economic Journal, Journal of Macroeconomics, Public Interest,* and *Challenge.*

Economic Parables & Policies